MTI Publication No. T-3

The Development of Controlled Hydrodynamic Techniques for Corrosion Testing

by

Tzu-Yu Chen, A. Moccari,
and D.D. Macdonald

Fontana Corrosion Center
Department of Metallurgical Engineering
Ohio State University
Columbus, OH 43210

Published for The Materials Technology Institute
of the Chemical Process Industries, Inc.
by the National Association of Corrosion Engineers

ABOUT MTI

The Materials Technology Institute of the Chemical Process Industries, Inc. (MTI) is a unique, cooperative research organization representing private industry. Its goal is to conduct generic, nonproprietary studies of a practical nature on the deterioration of materials and equipment used in the process industries.

Through membership in MTI, companies can solve nonproprietary problems of major concern to the process industries, leverage research dollars by participating in the direction and results of MTI studies, capitalize on the expertise of member company representatives, and learn about MTI's accomplishments in a timely fashion.

More information about membership in MTI can be obtained from the Executive Director. MTI members receive copies of MTI publications as part of their membership.

The Materials Technology Institute
of the Chemical Process Industries, Inc.
1215 Fern Ridge Parkway
Suite 116
St. Louis, MO 63141-4401
314/576-7712

CONTENTS

vi

LIST OF
TABLES

x

ACKNOWLEDGMENTS

The authors wish to recognize the contributions of G.D. Wu and Adil Kahan to the early research on the project.

The support and direction of A.M. Hall and E.A. Kachik of the Materials Technology Institute of the Chemical Process Industries, Inc. are also gratefully acknowledged.

MTI PERSPECTIVE

It has been recognized for many years that hydrodynamic effects are often important in determining the rate of corrosive attack on metals. Several approaches have been used by investigators to obtain some assessment, either qualitative or quantitative, of the magnitude of these hydrodynamic effects. These have included techniques such as pumping the corrosive fluid through tubular specimens, impinging the corrosive fluid onto a stationary specimen, rotating the specimens in a bath of corrosive fluid, or using gas bubbles to agitate the corrosive fluid in the vicinity of the specimens. The National Association of Corrosion Engineers (NACE) has published a standard test method for conducting controlled velocity laboratory corrosion tests, TM0270, "Method of Conducting Controlled Velocity Laboratory Corrosion Tests." However, these tests have generally involved complicated equipment and have often been less than successful in developing results that can easily be interpreted and applied in other system geometries.

The Technical Advisory Council of the Materials Technology Institute of the Chemical Process Industries, Inc. (MTI) had three objectives in view when it recommended the approval of Project 15, "The Development of Controlled Hydrodynamic Techniques for Corrosion Testing." The primary objective was to sponsor a fundamental study on the effects of hydrodynamic conditions on the kinetics of corrosion processes. The second objective was to seek out several approaches to the quantitative assessment of the hydrodynamic effects on corrosion kinetics. The third objective was to find, if possible, an approach to laboratory scale testing that would successfully simulate hydrodynamic conditions that exist in process equipment.

The Fontana Corrosion Center of the Ohio State University was selected for this project based on an unsolicited proposal for the following reasons. The investigators had a proven track record in studies on hydrodynamic effects. They were also judged to be knowledgeable and competent in the use of modern electrochemical techniques of corrosion rate measurement. This was an important aspect of the study because these techniques offer opportunities for rapid rate measurements, as well as provide insight into the mechanisms of the corrosion reactions.

The results of this study are presented in the following report. CDA Alloy 706, a 90:10 cupronickel alloy with about 1.5% iron in solid solution, was selected as the metal. The corrosive environment was a 1 M NaCl solution at 25°C (77°F). The pH and oxygen content of the solution were varied along with the hydrodynamic conditions. Three geometries were selected: pipe flow, annular flow, and rotating cylinder. The corrosion rates were measured by three different techniques: small amplitude cyclic voltammetry (SACV), alternating current (AC) impedance measurement, and mass loss rate as measured by weight loss. The electrochemical techniques are based on the relationship first found by Stern and Geary between polarization resistance and corrosion rate.

Geometries were found where the corrosion rate was determined by the hydrodynamic conditions. Consequently, the published correlations for predicting convective/diffusion rates for these flow geometries would be successful in predicting the corrosion rates. In general, the correlation between the pipe flow and the rotating cylinder was better than the annular flow channel. Thus, the rotating cylinder geometry has merit as an approach to simulating pipe flow. This is significant because a rotating cylinder apparatus is much more practical as a bench-scale laboratory technique than is a flow loop.

The SACV technique was judged to be better than AC impedance for estimating the corrosion rate. The system used in this study was not as simple as was initially anticipated. As a result, some of the interpretations presented in this report are not proven and therefore are speculative. A major problem is the fact that the valence of copper was assumed to be 1 even though the majority of copper corrosion products were found in the +2 valence. Several other problems were encountered, which makes some of the conclusions controversial. More work will be necessary in order to resolve these issues. Nevertheless, this report marks an important contribution to our understanding of hydrodynamic factors in corrosion processes.

The work of the task group in overseeing and reviewing this project has been especially helpful and, in particular, the contribution of D.C. Silverman.

EXECUTIVE SUMMARY

A detailed study has been carried out on the effect of hydrodynamic variables (flow rate and geometry), oxygen concentration, and solution pH on the corrosion of 90:10 Cu-Ni alloy in a flowing 1 M NaCl solution at 25°C (77°F). Three hydrodynamic geometries were investigated: a flow-through pipe system, an annular flow channel, and a rotating cylinder system. Flow rates employed in this study spanned the range from laminar flow to highly turbulent conditions, and all hydrodynamic states were carefully characterized in terms of the Reynolds (Re) and Schmidt (Sc) numbers. Corrosion rates have been measured using small amplitude cyclic voltammetry (SACV), alternating current (AC) impedance techniques, and direct weight loss determinations. Corrosion potentials have also been measured as a function of time, flow velocity, and solution composition (oxygen concentration, pH).

The principal objectives of this work were as follows:

- Design and construction of low-cost controlled hydrodynamic corrosion testing rigs for use in the laboratory for quantitative simulation of the effect of fluid flow on materials degradation in systems of different geometries.

- Development and testing of corrosion/mass transport correlations that will permit the transfer of corrosion data from one geometry to another and that will also allow the accurate simulation of in-plant corrosion phenomena in the laboratory.

The experimental findings of this work can be summarized as follows:

The Effect of Hydrodynamics on the Corrosion Rate

- For all the hydrodynamic systems employed and each Re number studied, the corrosion potential, as well as the rate of corrosion, increased with exposure time in the first few hours of exposure but subsequently decreased to a nearly constant value for long exposure times. The averaged steady state values of corrosion potential and corrosion rate increased with increasing Re numbers.

- Good agreement was obtained between the SACV and AC impedance techniques for measuring polarization resistance, and the calculated weight losses were in reasonable accord with directly measured values.

- The mass transfer rate correlations for the hydrodynamic systems studied are in general agreement with published data.

- The corrosion rate measured in all the hydrodynamic systems increases linearly with increasing Sherwood numbers (Sh_1 and Sh_2) and mass transfer coefficients (K_1 and K_2) on a log-log scale. This reflects that the rate of corrosion is controlled by the rate of mass transfer.

- In the region of high Re numbers ($Re > Re_{critical} = 10,000$), transfer of the corrosion rate from one geometrical system to another can be made based on the mass transfer coefficient for the diffusion boundary layer, K_h, calculated from a previously established mass transfer rate equation. The transfer of the corrosion rate from one geometry to another can also be achieved based on the overall mass transfer coefficient K_1; however, the relation between the mass transfer coefficient in the surface film, K_f, and the Re number for each hydrodynamic system must be experimentally determined.

The Effect of Dissolved Oxygen Concentration on the Corrosion Rate

- The corrosion rate and the corrosion potential measured in the flow-through pipe channel and the annular flow channel first increased but then decreased with increasing dissolved oxygen concentration.

- Good agreement was obtained between the SACV and AC impedance techniques for the measurement of polarization resistance, and the calculated weight losses were in reasonable accord with the values directly measured.

The Effect of pH on Corrosion Rate

- For the flow-through pipe channel and the annular flow channel, the averaged steady state corrosion potential first decreased with increasing pH but then became independent of it at $pH > 7$.

- The rate of corrosion measured in the flow-through pipe channel and the annular flow channel first decreased with increasing pH but then tended toward a constant value at $pH > 6$.

- Good agreement was obtained between the SACV and AC impedance techniques for the measurement of polarization resistance at $pH = 8.5$, and the calculated weight losses were in reasonable accord with directly measured values.

The technological implications of the present work have been discussed briefly at the end of the report with emphasis on how the systems employed in this study might be used to develop more effective corrosion testing techniques for use in the laboratory.

INTRODUCTION

A large amount of corrosion research is currently carried out in quiescent systems. Accordingly, hydrodynamic factors are ignored in the analysis of the kinetics of materials degradation, or at best, a hydrodynamic regime is employed that does not effectively simulate that found in an industrial environment. However, few industrial systems are quiescent; indeed, many involve the flow of extremely corrosive fluids through pipes, channels, etc., at high velocities. In these cases, mass transfer and surface shear effects may have a profound effect on the rate of material degradation, either by modifying the rate of transport of chemical species to or from the surface or by shear-stripping protective films from the alloy/solution interface. As a result, an accurate simulation of corrosion phenomena that occur in the plant environments can be made in the laboratory only if the hydrodynamic effects are taken into account.

The objectives of this work were as follows:

(1)   Design and construct low-cost controlled hydrodynamic corrosion testing rigs for use in the laboratory for quantitative simulation of the effect of fluid flow on materials degradation in systems of different geometries.

(2) Develop and test corrosion/mass transport correlations that will permit the transfer of corrosion data from one geometry to another and that will also allow the accurate simulation of in-plant corrosion phenomena in the laboratory.

(3) Investigate the viability of the methods developed in (1) and (2) by analyzing the corrosion behavior of 90:10 Cu-Ni heat exchanger tubing in well-defined geometries, such as the flow-through pipe channel, the annular flow channel, and the rotating cylinder system, under a variety of hydrodynamic and mass transfer regimes (laminar, transitionary, and turbulent flow). These studies have employed electrochemical methods (small amplitude cyclic voltammetry [SACV] and alternating current [AC] impedance techniques), as well as direct weight loss measurement of corrosion rates.

This report contains four chapters. Chapter One gives a brief review of the literature related to this work. Chapter Two describes the experimental facilities and procedures employed. Chapter Three discusses the experimental results, and the conclusions are presented in Chapter Four.

CHAPTER
ONE

LITERATURE REVIEW

1.1 EFFECT OF FLUID VELOCITY ON THE CORROSION RATE OF METALS IN CONDENSED MEDIA

The corrosion of metals and alloys in condensed media involves at least three fundamental steps:

(1) transport of the reactant to the metal surface,

(2) electron exchange resulting in metal loss, and

(3) transport of the corrosion products from the surface to the bulk solution.

Because these three steps occur in series, any one of them can control the rate of the overall process. Accordingly, it is necessary to take into account both charge transfer and mass transfer phenomena when describing the corrosion of metals and alloys in aqueous solutions. The corrosion process is under activation control when (2) is the rate-limiting step; whereas, it is under diffusion control when (1) or (3) becomes the rate-limiting step.

Few systematic studies report on the effect of velocity on the rate of the corrosion of metals and alloys in condensed media, particularly over a wide range of fluid velocities and on a comparative basis using different flow geometries. In general, the obtained data schematically follow the trends shown in Figure 1.

At low velocities (Region I), the rate of corrosion is totally or partly controlled by the rate of mass transfer of a corrosive species to, or a corrosion product from, the surface. The effect of increasing the velocity on the rate of corrosion is to increase the surface concentration of the reactant or to decrease the interfacial concentration of the product. Therefore, the corrosion rate increases with fluid velocity. At higher velocities, the rate of mass transfer is much greater than the rate of the surface (corrosion) reaction, and activation control of the overall process is observed. In this case, the corrosion rate is independent of the fluid velocity. However, at still higher flow velocities, shear forces strip the protective films from the surface and give rise to an enhanced corrosion rate due to erosion-corrosion.

1.2 THEORY OF DIFFUSION-CONTROLLED CORROSION REACTIONS

The transport of reactive species between the bulk solution and the electrode/solution interface can occur by diffusion due to a concentration gradient and/or by migration due to a potential field. The steady rate of discharged species in a stagnant solution can be expressed as follows:

$$\frac{i_\ell}{nFA} = \pm D\frac{dC}{dy} \pm Cu\frac{d\phi}{dy} \tag{1}$$

in which i_ℓ is the limiting current, n is the number of electrons involved in the reaction, F is the Faraday's constant, A is the electrode surface area, D is the diffusivity of the diffusion-limited species, dC/dy is the concentration gradient normal to the electrode surface, C is the concentration, u is the mobility, and dϕ/dy is the potential gradient normal to the electrode surface. The migration term ($\pm$ Cu dϕ/dy, Equation [1]) can be neglected when an excess of supporting electrolyte is used in the solution. The integration of Equation (1) then gives:

$$\frac{i_\ell}{nFA} = \pm\frac{D}{\delta}(C-C_o) = \pm K(C-C_o) \tag{2}$$

in which δ is the thickness of the diffusion boundary layer, C is the bulk concentration, C_o is the surface concentration, and K is the mass transfer coefficient. When the limiting current value is reached, C_o becomes zero, and the corrosion rate is controlled by the diffusion of the cathodic reactant to the metal surface, or if C_o is equal to the saturation concentration of the corrosion product, the corrosion rate is controlled by the transport of the anodic corrosion product away from the metal surface.

Most of the work studying the effect of flow on corrosion was based on fluid velocity. However, several authors emphasized the importance of using hydrodynamic and/or mass transfer parameters in analyzing corrosion behavior so that the corrosion data obtained in the laboratory can be used with more confidence in field applications.[1-3]

For a given hydrodynamic or geometrical system, the mass transfer correlation can be expressed in terms of Sh, Re, and Sc in the following form:

$$Sh = A\,Re^{m}Sc^{n} \tag{3}$$

in which $Sh = KX/d$, $Re = \rho XV/\mu$, and $Sc = \mu/\rho D$ are Sherwood, Reynolds, and Schmidt numbers, respectively. X is the characteristic dimension, ρ is the fluid density, μ is the viscosity, and D is the species diffusivity. The constants A, m, and n in the equation are generally obtained by theoretical analysis or empirically by experimentation.[4-6] Experimentally, they are usually determined by measuring the limiting current using electrochemical methods.[7] Typically, $1/2 \geq m \leq 1$ and $1/4 \leq n \leq 1$.

From Equation (2), for a diffusion-controlled corrosion reaction, the rate of corrosion can be determined by the limiting current density and is given by:

$$\text{corrosion current} = \frac{i_{\ell}}{A} = nFK\Delta C \tag{4}$$

and substitution for K in Equation (4) then gives:

$$\text{corrosion rate} = nF\Delta C \left(\frac{D}{X}\right) Re^{m}Sc^{n} \tag{5}$$

For a given hydrodynamic system, the corrosion rate can then be determined without testing by inserting the appropriate values of the relevant parameters in the above equation but only if the reaction is mass transfer controlled. However, natural convection should be considered at zero to low Re numbers because Equation (5) would predict that no corrosion occurs at Re = 0.[8]

1.3 REVIEW OF MASS TRANSFER CORRELATIONS

B. Poulson, in an excellent review, described existing laboratory techniques for studying the effects of flow velocity on electrochemical and corrosion processes.[8] The practical and mechanistic significance of electrochemical measurements in flowing solutions was also discussed. The methods adopted in the laboratory for simulating fluid flow effects on corrosion should satisfy a number of basic requirements:

- The hydrodynamic laminar, transitionary, and turbulent regimes and flow characteristics should be well-defined.
- Mass transfer correlations should be readily available or should at least be easily determined.
- The corroding surface should be equipotential so that the material corrodes at the same rate and by the same mechanism at all points.

Consideration of hydrodynamic and mass transport correlations obtained from a large number of studies indicates that the most suitable geometries satisfying the above-mentioned requirements are the flow-through pipe, the annular flow, and the rotating cylinder systems. The first two are appropriate only if a suitable approach section is used to fully develop both the hydrodynamic and diffusion boundary layers ahead of the specimen. The frequently used rotating disk configuration is rejected because a number of studies have shown that the surface is not equipotential except under complete mass transfer control. Furthermore, the Re number at the surface varies from zero at the disk center to a maximum at the edge of the disk. Both factors can lead to difficulties in generating a well-defined hydrodynamic/corrosion regime using this geometry. For this reason, mass transfer correlations of the flow-through pipe channel, the annular flow channel, and the rotating cylinder system studied in this work are reviewed in this chapter.

1.3.1   The Flow-Through Pipe Channel

This type of flow system is probably the most commonly used geometry for the transmission of fluids. The counter electrode can be placed either upstream, downstream, or at both locations with respect to the tubular working electrode.

1.3.1.1 *Laminar Flow Regime*

Blaedal et al. studied the oxidation of ferrocyanide on platinum electrodes of radii from 0.254 to 0.508 mm.[9] They reported that the electrolysis current is proportional to a 0.335 power of the flow rate and claimed that their results were in good agreement with the Leveque equation:

$$I = 5.24 \cdot 10^5 n C_b D^{\frac{2}{3}} X^{\frac{2}{3}} V^{\frac{1}{3}} \tag{6}$$

in which X is the tube diameter and V is the flow rate.

Sharma and Dutt used a tubular graphite electrode of 0.05-cm radius and 2.5 cm in length to measure the electrolysis current.[10] They reported that the electrolysis current was proportional to a 0.36 power of the flow rate, which was slightly higher than the theoretical value of 0.33. The discrepancy was attributed to local turbulence at the tube entrance. However, Devanathan and Guruswamy[11] have criticized Blaedal et al.[9] and Sharma and Dutt[10] for claiming that their results are in good agreement with the Leveque equation for tubes because the conditions assumed by Leveque in his analysis were not satisfied in their experiments.

1.3.1.2 *Turbulent Flow Regime*

The most common correlations for fully developed turbulent flow in smooth straight tubes are the Chilton-Colburn analogy:[12]

$$Sh = 0.023 Re^{0.8} \cdot Sc^{0.33} \tag{7}$$

and the Harriot-Hamilton correlation:[13]

$$Sh = 0.0096 Re^{0.913} Sc^{0.346} \tag{8}$$

B. Poulson summarized a large number of correlations for this flow condition as shown in Table 1.[8] Coney claimed that the Chilton-Colburn analogy would diverge at high Re numbers and give low Sh numbers at high values of Sc.[14] Therefore, the Berger and Hue correlation was recommended.[15]

1.3.2   **The Annular Flow Channel**

1.3.2.1 *Laminar Flow Regime*

Lin et al. correlated their experimental data for an annulus geometry of diameter ratio of $d_1/d_2 = 0.5$ in which the electrode of interest formed part of the inner cylinder while the outer cylinder formed the counter electrode.[16] They claimed that their correlation:

$$\frac{KX}{D} = 1.62\left(\frac{GX}{\rho D}\right)^{\frac{1}{3}}\left(\frac{X}{L}\right)^{\frac{1}{3}} \tag{9}$$

was in good agreement with Leveque's equation:

$$\frac{KX}{D} = Sh = 1.614\,(Re\,Sc)^{\frac{1}{3}}\left(\frac{X}{L}\right)^{\frac{1}{3}} \tag{10}$$

for a circular pipe in the laminar flow region. In Equation (9), G is the mass velocity (g/cm^2/s) and X is the effective diameter. Friend and Metzner criticized Lin's assumption that the Leveque equation for tubular flow may be applied to the annular geometry simply by replacing the diameter of the pipe with the effective diameter of the annulus.[17] They suggested that the extra variables, which the annulus geometry involves, requires an additional dimensionless group that is a function of d_1/d_2.

Ross and Wragg derived mass transfer correlations for fully developed laminar flow in annuli based on the Leveque solution for laminar flow passing a flat plate by modifying the velocity profile terms.[18] They obtained:

$$Sh = \frac{KX}{D} = 1.614[\phi(\alpha)]^{\frac{1}{3}}(Re\,Sc)^{\frac{1}{3}}\left(\frac{X}{L}\right)^{\frac{1}{3}} \qquad (11)$$

in which

$$\phi(\alpha) = \frac{1-\alpha}{\alpha}\,\frac{0.5 - \dfrac{\alpha^2}{1-\alpha^2}\ln\left(\dfrac{1}{\alpha}\right)}{\dfrac{1+\alpha^2}{1-\alpha^2}\ln\left(\dfrac{1}{\alpha}\right)-1} \qquad (12)$$

and

$$\alpha\,\frac{d_1}{d_2}$$

They used different annular diameter ratios in their experiments and obtained different correlations as follows:

for

$$\frac{d_1}{d_2} = 0.5, \quad Sh = 1.76\,(Re\,Sc)^{\frac{1}{3}}\left(\frac{X}{L}\right)^{\frac{1}{3}} \qquad (13)$$

for

$$\frac{d_1}{d_2} = 0.25, \quad Sh = 1.8\,(Re\,Sc)^{\frac{1}{3}}\left(\frac{X}{L}\right)^{\frac{1}{3}} \qquad (14)$$

and for

$$\frac{d_1}{d_2} = 0.125, \quad Sh = 2.02\,(Re\,Sc)^{\frac{1}{3}}\left(\frac{X}{L}\right)^{\frac{1}{3}} \tag{15}$$

A comparison of the results using Equation (13) ($d_1/d_2 = 0.5$, i.e., the diameter ratio used by Lin et al.[16]) with those of Lin and co-workers showed that they are below the theoretical value by an average of approximately nine percent, while the discrepancy of Lin's data is about 17 percent.

From Equations (13), (14), and (15), it can also be seen that the constant term in the correlations increases as the value of d_1/d_2 decreases. This demonstrates that the correlation derived by Lin et al., which merely substitutes the effective diameter of the annulus in the equation applicable to a tube, is inadequate.

1.3.2.2 Turbulent Flow Regime

Lin et al. reported that their experimental data were in good agreement with the Chilton-Colburn correlation for the flow region 2,100 < Re < 30,000 and 325 < Sc < 3,110.[16] However, Friend and Metzner[17] pointed out that Lin's data were about 22 percent lower than the empirical relationship of Monard and Pelton[19] for turbulent flow in an annulus.

Ross and Wragg[18] derived the mass transfer correlation for turbulent flow in an annulus based on the differential equation for diffusion and convection in the system following the procedure of Linton and Sherwood[20] and assuming that the velocity distribution in the region close to the wall is approximately linear. Their experimental correlation, in terms of Stanton's number ($St = Sh/Re/Sc$), given as:

$$St = 0.276\,Re^{-0.42}\,Sc^{-\frac{2}{3}}\left(\frac{X}{L}\right)^{\frac{1}{3}} \qquad (16)$$

in which the tube diameter has been substituted for the effective diameter, is about nine percent below the Chilton-Colburn correlation.

1.3.3   The Rotating Cylinder System

The rotating cylinder has been used extensively in corrosion research not only because of the ease and low cost of its fabrication but also because of the fact that the flow is essentially turbulent even at very low rotational velocities.

1.3.3.1 *Laminar Flow Regime*

Cornet and Kappesser examined the mass transfer behavior for one solution ($Sc = 46$) over a wide range of Re numbers.[21] The critical Re number for the transition from laminar to turbulent flow was found to be approximately 200. In the laminar flow regime ($Re < 200$), the Sh number was found to be constant ($Sh = 37$) and was independent of Re. They suggested that the fluid is rotating with the inner rotating cylinder without any significant contribution to the enhancement of the rate of mass transfer. For turbulent flow ($Re > 200$), the mass transfer rate was correlated as:

$$Sh = 0.97\,Re^{0.64} \qquad (17)$$

Gabe[22] derived the mass transfer correlation for laminar flow by considering the transport behavior at a segment length of the inner rotating cylinder and by using the velocity profile equation for the annulus ($R^2 - R_1$) developed by Levich[23] and Bird et al.[24] He obtained:

$$Sh = \frac{0.64\,Re\,Sc\,F(R)}{2\pi L} \qquad (18)$$

$$F(R) = \left[\frac{\left(1 + \dfrac{R_1^2}{R_2^2}\right)}{\left(1 - \dfrac{R_1^2}{R_2^2}\right)} \right]^{\frac{1}{3}} \qquad (19)$$

Kimla and Strafelda obtained a different correlation using a somewhat different velocity profile (Equation [20]);[25] however, the dependence of Sh upon Re and Sc remains the same:

$$Sh = 0.38\left[\frac{Re\,Sc}{2\pi L}\left(\frac{1 - R_1^2}{R_2^2} \right) \right]^{\frac{1}{3}} \qquad (20)$$

1.3.3.2 *Turbulent Flow Regime*

For the case of a rotating inner cylinder and static outer cylinder, Eisenberg et al. obtained for the ferri-ferrocyanide redox reaction:[26]

$$Sh = 0.0791\,Re^{0.7}Sc^{0.356} \qquad (21)$$

They also found that the critical dimension was the diameter of the rotating cylinder rather than the gap. Robinson and Gabe confirmed the Eisenberg mass transfer correlation by a systematic study of the electrodeposition of copper from $CuSO_4$-H_2SO_4 solutions.[27,28]

Newman has pointed out that the ratio of inner to outer cylinder radii, R_1/R_2, should be incorporated in the mass transfer correlation and that its effect is to modify Re.[29,30] He proposed that a more appropriate correlation should read:

$$Sh = 0.0791\left[Re\left(\frac{R_1}{R_2}\right)\right]^{0.7} Sc^{0.356} \qquad (22)$$

The Eisenberg correlation is obtained when $R_1/R_2 \rightarrow 1$. Any departure from this condition will result in a change in the constant.

Kappesser et al. investigated the effect of surface roughness on mass transfer to rotating cylinders over a wide range of Re numbers for the surface roughness term d/ϵ (ratio of specimen diameter to roughness height) ranging from 87 to a smooth surface, and proposed the following correlation:[31]

$$Sh = \left(\frac{f}{z}\right) Re\, Sc^{0.356} \qquad (23)$$

in which f is the friction factor.

Coeuret and LeGrand investigated the mass transfer behavior at the electrodes of concentric cylindrical reactors combining axial flow and the rotation of the inner cylinder and proposed an empirical relationship of the type:[32]

$$Sh = a Re_a^b Ta^c Sc^d \qquad (24)$$

in which Re_a is the axial Reynolds number and Ta is the Taylor number $(Ta = [R_1\{R_2 - R_1\}\omega/v][\{R_2 - R_1\}/R_1]^{1/2}$ in which ω is the angular velocity of the electrode). They deduced that for small cylindrical gaps ($R_2 - R_1$ = 2.5 mm and $R_2 - R_1$ = 5.0 mm):

$$Sh = 0.12 Re_a^{\frac{1}{3}} Ta^{0.4} Sc^{\frac{1}{3}} \quad \text{for } Re_a > 300 \tag{25}$$

and

$$Sh = 0.38 Ta^{\frac{1}{2}} Sc^{\frac{1}{3}} \quad \text{for } Re_a < 300 \tag{26}$$

Many workers have paid little or no attention to the fluid dynamics of rotation and have concentrated only on achieving a specific rotating speed.[33,34] Gabe has provided a detailed discussion of the design, construction, and geometrical factors.[22]

TABLE 1

Some Correlations for Fully Developed Tube Flow[7]

Originator	Correlation	Origin and Comments
Chilton-Colburn	$Sh = 0.023\, Re^{0.8} Sc^{0.33}$	Analogy with heat transfer
Berger-Hue	$Sh = 2 + c Re^{a} Sc^{0.3}$ $c = 0.0165 + 0.011\, Sc\, e^{-Sc}$ $a = 0.86 - 10/(4.7 + Sc)^{3}$	Their own limiting current density and other experimental data
Berger-Hue	$Sh = 0.0165\, Re^{0.86} Sc^{0.33}$	For Sc = 10, only 3% error
Harriot-Hamilton	$Sh = 0.0096\, Re^{0.913} Sc^{0.346}$	From their chemical dissolution experiments involving benzoic acid in glycerine-water mixtures
Notter-Sleicher	$Sh = 0.0149\, Re^{0.88} Sc^{0.33}$	Theoretical diffusivity with constants from experimental work
Churchill	$Sh = Sh_0 + 0.079 \sqrt{\dfrac{f}{2}}\; \dfrac{ReSc}{(1 + Sc^{0.8})^{\frac{5}{16}}}$	Empirical analogy with friction
Jayattilleke	$Sh = \dfrac{ReSc}{0.9\left[P\sqrt{\dfrac{2}{f}} + \dfrac{2}{p} = 7.81 \right]}$ where $p = f(Sc)$	Boundary layer integration; total resistance to mass Transfer is the sum of resistances in laminar and turbulent regions

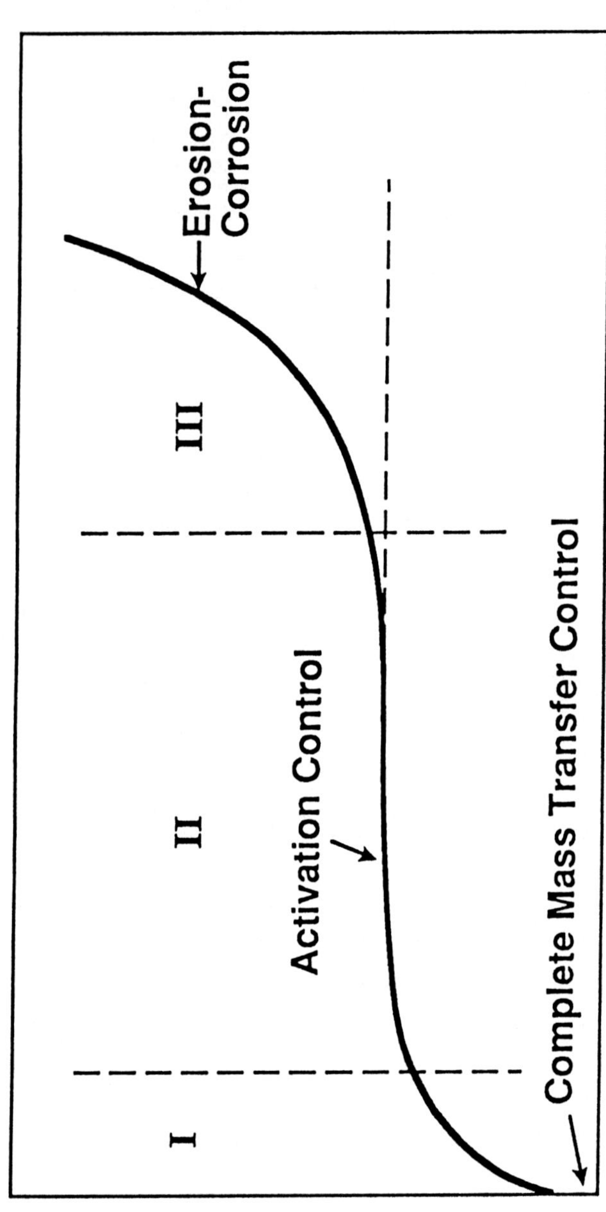

Figure 1   Schematic Variation of Corrosion Rate with Fluid Velocity

TWO

EXPERIMENTAL DETAILS

A schematic diagram of the flow loop is shown in Figure 2. The flow loop was constructed from 1-in. internal diameter (ID) polyvinyl chloride (PVC) pipe (Schedule 80) and contained a 30-gallon polyethylene reservoir with facilities for gas sparging, a glass-filled epoxy plastic centrifugal pump, bypass lines for flow velocity control, a flow meter with a paddle wheel flow sensor, a pH controller, and a portable refrigerator. Flow velocities of 0 to 4.5 m/s could be obtained. The apparatus was designed to accommodate several test sections (a flow-through pipe channel, an annular flow channel, and a rotating cylinder system) simultaneously.

2.1 TEST SECTION

The specimens used in this study were fabricated from tubular 90:10 Cu-Ni alloy with the chemical composition and dimensions as summarized in Table 2. Specimens were used in their as-received condition and were rinsed in acetone and then air-dried prior to exposure.

2.1.1   <u>The Flow-Through Pipe Channel and the Annular Flow Channel</u>

Figure 3 shows the specimen flow channels that were used in this work. The test sections comprised five tubular specimens separated from each other by thin Delrin spacers. Each set of five specimens was preceded by a 75-cm long approach section of Cu-Ni tubing of identical composition and internal diameter as the specimens to ensure full development of the mass transfer and velocity boundary layers upstream from the specimens. Accordingly, all specimens were tested under essentially identical conditions. In each set of five specimens, counted in a direction from upstream to downstream, the first and fifth specimens were used for measuring polarization resistance using the small amplitude cyclic voltammetry (SACV) method, the second specimen for alternating current (AC) impedance studies, the third specimen for gravimetric corrosion rate measurements using a microbalance, and the fourth specimen was used for recording the variation of corrosion potential with exposure time, as well as for potentiodynamic polarization studies. Because the mass transfer conditions were nominally identical for all five specimens in each test channel, the corrosion rate obtained for one specimen was expected to be entirely consistent with that for any other specimen in that test channel. All potentials were measured against a saturated calomel electrode (SCE), which was supported in a side arm, just downstream from each set of five specimens. Both the approach section and the tubing just downstream from the test section (~15-cm long) were used as the counter electrode.

2.1.2   The Rotating Cylinder System

Figure 4 shows the design of the rotating cylinder system. Four specimens, separated from each other by thin Delrin spacers, were supported on a heavy-wall 304 stainless steel (UNS S30400) tubing shaft. For the specimens counted from top to bottom, the first specimen was used for SACV measurements, the second specimen was used for recording the variation of corrosion potential with time and for potentiodynamic polarization studies, the third specimen was used for AC impedance measurements, and the fourth specimen was used for direct weight loss measurement using a microbalance. A concentric stationary outer copper tube (inside diameter = 3.81 cm) was used as the counter electrode. Three holes (1.905 cm in diameter) were bored on both the top side surface and the bottom side surface of the counter electrode to serve as the entrance for the solution into the gap between the outer stationary counter electrode and the inner rotating electrodes. All potential measurements were made with respect to an SCE through a Luggin probe that was inserted into the gap through one hole on the top side surface of the counter electrode. The rotating cylinder system was connected to the flow loop via a polyethylene tube. The low solution refreshing rate (~50 cc/min) ensured that the effect of axial flow on the mass transfer rate correlation and/or the rate of corrosion was negligible.

Carbon brushes were found to be satisfactory for electrical contact with the rotating shaft.

For the experiment conducted at 1,000 rpm (Re = 18,040), a graphite impregnated polytetrafluoroethylene (PTFE) O-ring was used to seal the test cell with the solution flowing back into the reservoir through outlet No. 1 (Figure 4). However, after about 40 hours of exposure, the ball bearing in the bottom side of the aluminum block was found to be worn out. In this case, all measurements were then made on Specimen 1, where vibration of the shaft was less severe, in the order of SACV, AC impedance, and then polarization studies at time intervals of 24 hours. The PTFE O-ring was also found to be worn out after about 100 hours. A change in the design of the rotating cylinder system was then made for the experiments performed at 2,000 rpm and 600 rpm, and acrylic tubing (outer diameter [OD] = 0.635 cm) was used to serve as the solution outlet (Solution Outlet No. 2 in Figure 4). As soon as the solution level reached the top edge of the acrylic tube, the solution passed into a polyethylene tank located below the test cell. The solution stored in this tank was then pumped back into the reservoir using a pump with an automatic on-off timer. The shaft was driven by a DC motor with the rotation speed adjusted by a speed controller. A magnetic sensor was used as pick-up for the digital read-out on a tachometer. The rotation speed was controlled to within ±15 rpm of the desired value.

2.2 MEASUREMENT OF DISSOLVED OXYGEN CONCENTRATION

In order to study the effect of dissolved oxygen concentration on the rate of corrosion, the reservoir was continuously purged with the desired oxygen-nitrogen gas mixture during each run, except for the experiments conducted under air-saturated conditions, in which case severe agitation of the solution in the reservoir in contact with ambient air ensured air saturation. The dissolved oxygen concentration was measured using a colorimetric method. The dissolved oxygen content in air-saturated 1 M NaCl solution was found to be about 8 ppm.

After the experiment, the dissolved oxygen content of the solution was also measured using an oxygen sensor and an oxygen analyzer. The oxygen analyzer was first calibrated with air-saturated double distilled water ($[O_2]$ ~8.4 ppm). The dissolved oxygen content in air-saturated 1 M NaCl solution (pH = 5) was then found to be approximately 8.1 ppm.

2.3 CONTROL OF THE SOLUTION pH

A pH controller and a PTFE solenoid valve were incorporated into the system to control the pH of the solution. A dilute (~4%) HCl solution was used to adjust the pH of the solution to within ±0.1 for those experiments conducted at pH = 5, 6, and 7. For the experiment performed at pH = 8.5, a 4 N NaOH solution was used to adjust the pH to within ±0.2 of the desired value.

2.4 CONTROL OF THE SOLUTION TEMPERATURE

An increase in the solution temperature from 25°C (77°F) to approximately 40°C (104°F) was experienced during the initial period of this work due to pumping work. The problem was solved by placing a copper coil in the reservoir through which water cooled by a portable refrigeration coil was passed. The temperature of the solution was then controlled to 25 ± 2°C (77 ± 4°F).

2.5 CORROSION RATE MEASUREMENTS

2.5.1   Small Amplitude Cyclic Voltammetry (SACV)

SACV data were obtained by imposing a small amplitude (20-mV peak-to-peak) triangular potential excitation across the interface. Voltage sweep rates over the range 0.1 to 50 mV/s were employed, and the current/voltage curves were recorded with an X-Y recorder.

2.5.2   Alternating Current (AC) Impedance

AC impedance data were generated as a function of frequency (10^4 Hz to 10^{-2} Hz or lower) by imposing a 20-mV peak-to-peak sinusoidal voltage signal across the interface. The complex impedance data were acquired using a frequency response analyzer controlled by a microcomputer.

2.5.3   Direct Weight Loss Measurements

The specimens (Sample No. 3 for the flow-through pipe channel and the annular flow channel and Sample No. 4 for the rotating cylinder system) were rinsed in acetone, air-dried, and then weighed prior to exposure. After exposure, the specimens were descaled by immersion in a 10% HCl solution, scrubbed, rinsed in distilled water, rinsed in acetone, air-dried, and then reweighed in order to calculate the weight losses due to corrosion.

2.6 POTENTIODYNAMIC POLARIZATION MEASUREMENTS

Potentiodynamic polarization curves were obtained by first sweeping the voltage at a scan rate of either 25 mV/min or 30 mV/min from the corrosion potential into the cathodic region and then into the anodic region. The polarization curves were obtained using a potentiostat and a motor potentiometer.

TABLE 2

**Chemical Composition and Dimensions
of the 90:10 Cu-Ni Tubular Specimen Used in This Work**

Element	Cu	Ni	Fe	Mn	Pb	Zn
Wt %	88.45	9.40	1.35	0.65	0.02	0.01

Dimensions of tubular specimen:
Inside Diameter = 1.575 cm
Outside Diameter = 1.905 cm
Length = 1.270 cm

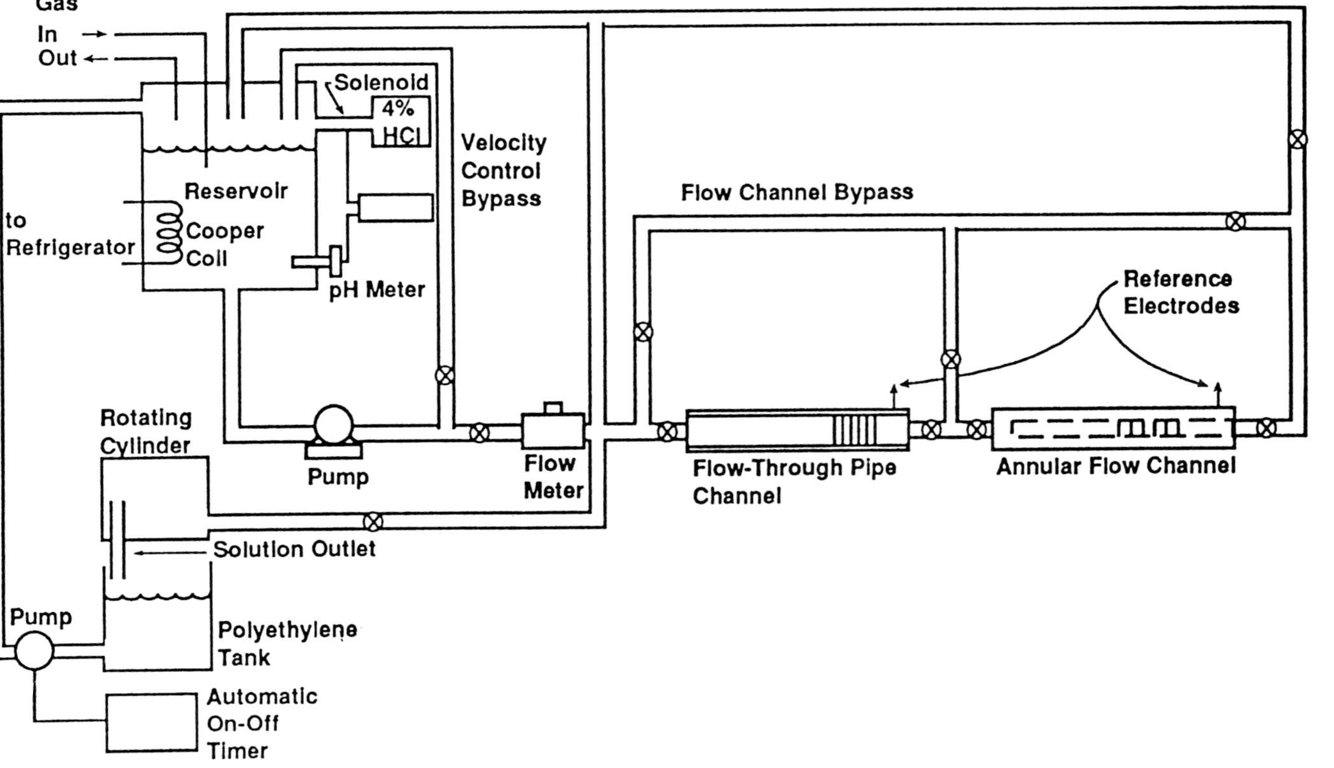

Figure 2   Schematic Diagram of the Flow Loop

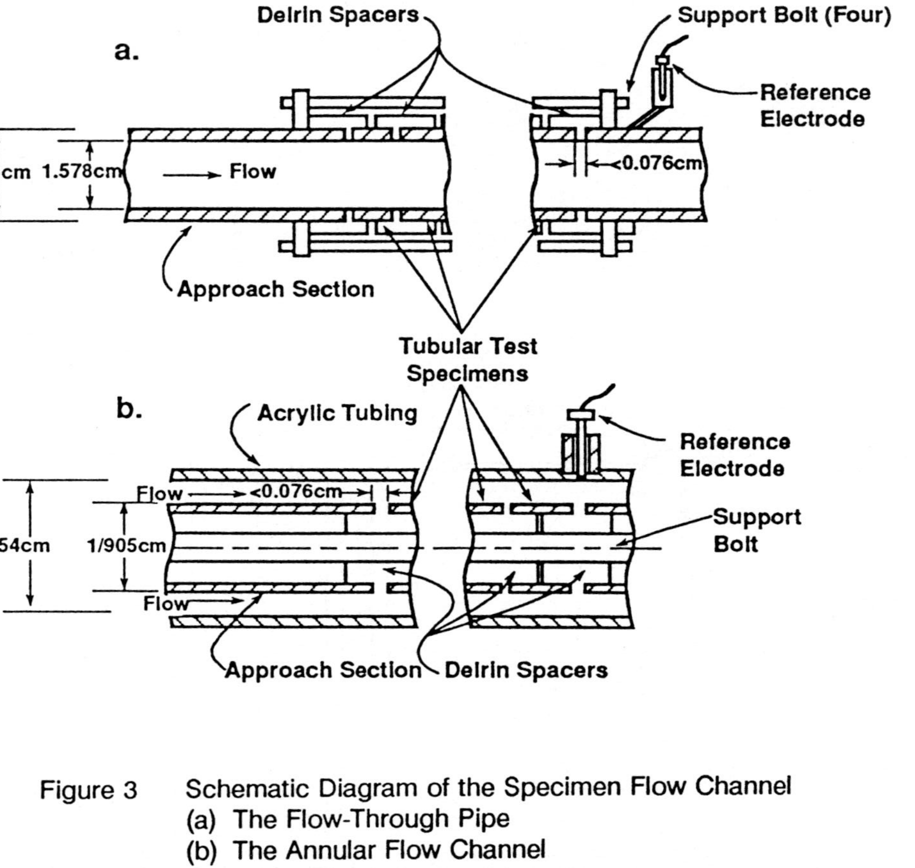

Figure 3   Schematic Diagram of the Specimen Flow Channel
(a) The Flow-Through Pipe
(b) The Annular Flow Channel

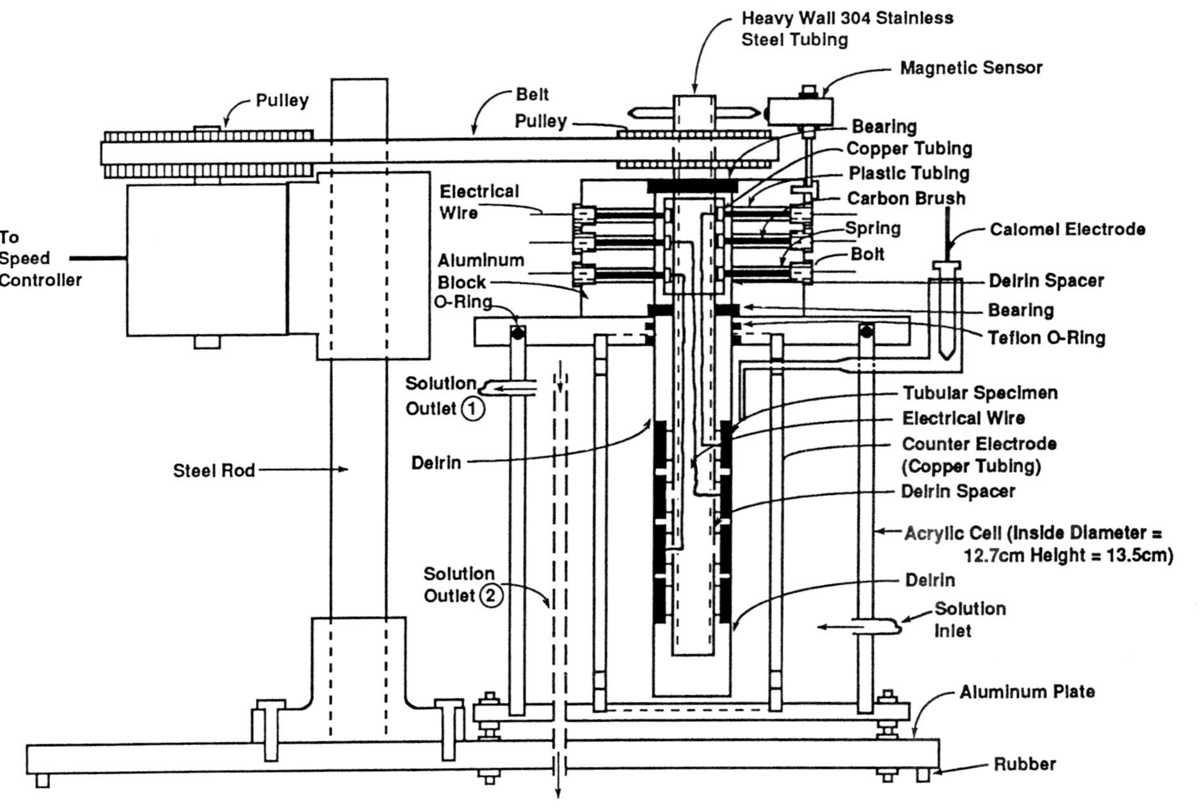

Figure 4   Schematic Diagram of the Rotating Cylinder System

THREE

RESULTS AND DISCUSSION

3.1 THE EFFECT OF HYDRODYNAMICS ON CORROSION

The effect of hydrodynamics on the corrosion of 90:10 Cu-Ni alloy in a flowing air-saturated 1 M NaCl solution (T = 25°C [77°F]) was studied at pH = 5 in systems of different geometries (the flow-through pipe channel, the annular flow channel, and the rotating cylinder system) under a variety of hydrodynamic and mass transfer conditions (laminar, transitionary, and turbulent flow).

3.1.1   Variation of Corrosion Potential ($E_{corrosion}$) with Time

Figures 5 through 7 show the variation of corrosion potential with exposure time at various Reynolds numbers (Re) for the flow-through pipe channel, for the annular flow channel, and for the rotating cylinder system, respectively. For each hydrodynamic system, at each Re number studied, the corrosion potential first increased with time during the first few hours of exposure but subsequently decreased to a nearly constant value for long exposure times. For each system studied, the steady state corrosion potential increased with increasing Re with only one exception—that for the rotating cylinder system at Re = 18,040. This is believed to be due to the wearing out of the ball bearings causing greater vibration of the rotating shaft.

The effect of increasing the Re number is to increase the availability of oxygen at the surface, which in turn will polarize the corrosion reaction in the more noble direction. Therefore, the steady state corrosion potential is expected to increase with increasing Re, as observed. The increase in the corrosion potential in the initial period of exposure, at each Re studied, is probably due to the dissolution of an air-formed surface film that existed prior to exposure.

.1.2   <u>Variation of Corrosion Rate (as 1/R_p) with Time</u>

3.1.2.1 *Small Amplitude Cyclic Voltammetry (SACV)*

Typical small amplitude cyclic voltammograms (SACVs) for 90:10 Cu-Ni alloy in a flowing 1 M NaCl solution measured at various potential scan rates are shown in Figure 8. Hysteresis loops were always observed at very high-voltage scanning rates. However, at lower scan rates (1 mV/s to 0.1 mV/s), straight lines with constant slope were always obtained. In some cases, nonlinear responses were obtained. Macdonald analyzed the transient response of a simple equivalent electrical circuit for an electrode/solution interface to a small amplitude triangular potential excitation.[44] He showed that the apparent polarization resistance (R_{app}) obtained by SACV can vary greatly with the potential sweep rate, particularly when the interface has a large capacitive element. Macdonald also showed that the true polarization resistance (R_p), which is inversely proportional to the rate of corrosion, could be obtained by extrapolating the curve of $1/R_{app}$ vs potential sweep rate to zero sweep rate. In addition, his analysis showed that $R_{app} \approx R_p$ when the cyclic voltammogram recorded showed negligible hysteresis. In this work, $R_{app} \approx R_p$ at a sufficiently low potential sweep rate (e.g., 0.1 mV/s); therefore, only these polarization resistance values are reported below.

Figures 9 through 13, Figures 14 through 18, and Figures 19 through 21 show the variation of corrosion rate (as 1/R_p) measured by SACV as a function of exposure time for the flow-through pipe channel, for the annular flow channel, and for the rotating cylinder system, respectively. For the flow-through pipe channel and the annular flow channel, average values of $1/R_p$ obtained from SACV on Specimen 1 and Specimen 5 are used. For each system, at each Re, the corrosion rate (as 1/R_p)

increased in the initial few hours of exposure but subsequently decreased to a nearly constant value over the exposure times employed in this study. This behavior parallels the variation of the corrosion potential with time, and the initial increase in the corrosion rate is also probably due to the dissolution of the air-formed oxide film.

3.1.2.2 Alternating Current (AC) Impedance

Alternating current (AC) impedance measurements were performed at various intervals during each run after the first 24 hours of exposure. Figures 22 through 26, Figures 27 through 31, and Figures 32 through 34 show the complex plane impedance diagrams at various Re numbers and exposure times for the flow-through pipe channel, the annular flow channel, and the rotating cylinder system, respectively. Macdonald has emphasized that valid R_p values can be obtained using the AC impedance technique only if the low-frequency locus in the complex plane impedance diagram is essentially complete.[35]

In the present work, excitation frequencies down to 10^{-2} Hz (or lower) were used, but in some instances (Figures 22, 24, 27, and 29), no clear intercept was found either in the high-frequency region or in the low-frequency region. Because these data do not allow a reasonable estimate of R_p to be made, they had to be discarded. The high-frequency points lying below the real axis (Figures 27 and 29) are due to the reactive properties of the current-measuring resistor (100 ohms) in series with the counter electrode. The inductive component disappeared when the current measuring resistor was replaced by one with a lower resistance (1 ohm). Inductance loops were observed below the real axis in the low-frequency region of the complex plane impedance diagram at high Re numbers for the flow-through pipe channel and the annular flow channel (Figures 24 through 26, 30, and 31). However, they were not observed for the rotating cylinder system at all the Re numbers studied for frequencies as low as 10^{-2} Hz.

The points of intercept of the locus (or extrapolated locus) in the high-frequency region and in the low-frequency region were taken to be R_s (the solution resistance) and $R_s + R_p$ (sum of the solution resistance and the polarization resistance), respectively. The value of R_p was then obtained by subtracting the high-frequency intercept (R_s) from the low-frequency intercept ($R_s + R_p$).

Values for $1/R_p$ for each hydrodynamic system were plotted as a function of exposure time. They are included in Figures 9 through 21, whenever values of $1/R_p$ could be reasonably estimated. Good agreement was obtained between the SACV and AC impedance techniques for measuring the polarization resistance. Values for $1/R_p$, obtained by SACV for exposure times when almost constant values of $1/R_p$ were obtained, were averaged for each Re number and each hydrodynamic system. The averaged values of $1/R_p$ were then normalized by dividing by the exposure area (A) and plotted as a function of velocity (linear flow velocity [V] for the flow-through pipe channel and the annular flow channel, or peripheral velocity [U] for the rotating cylinder system) and Re number (Figures 35 and 36, respectively).

For each hydrodynamic system, the corrosion rate (as $1/R_pA$) increases with increasing Re number (and velocity) over the range of Re numbers studied. This reflects the fact that the rate of corrosion is controlled, at least in part, by the rate of mass transfer. In this case, the effect of increasing Re number (or velocity) on the rate of corrosion is to increase the interfacial concentration of the reactant (oxygen).

Therefore, the corrosion rate increases with increasing Re number (and velocity).

3.1.3   <u>Comparison Between Measured and Calculated Weight Loss</u>

The corrosion rate data shown in Figures 9 through 21 are given in terms of $1/R_p$. Stern and Geary have shown that $1/R_p$ is directly proportional to the corrosion current (i_{corr}) as:[36]

$$i_{corr} = \frac{AB_aB_c}{2.303(B_a + B_c)R_p} \tag{27}$$

in which A is the area of the specimen, and B_a and B_c are the anodic and cathodic Tafel slopes, respectively.

Syrett and Macdonald developed an equation for deriving weight loss data from $1/R_p$ values obtained by SACV or other electrochemical techniques.[37] For 90:10 Cu-Ni alloy in seawater at a flow velocity of 1.62 m/s, they found:

$$\Delta W = 2.563 \cdot 10^{-4}BA_t \tag{28}$$

in which ΔW is the weight loss in units of grams, $B = B_aB_c/(B_a+B_c)$ in units of volts, and $A_t = \int_0^t 1/R_p \, dt$ in units of coulombs/volts.

A_t is the area under the $1/R_p$ (obtained by SACV) vs time curve. The values of A_t calculated by graphic integration of the curves shown in Figures 9 through 21 are listed in Table 3. Syrett and Macdonald also calculated weight loss values by assuming $B = 0.07$ volt ($B_a = 0.07$ volt and $B_c = \infty$) in Equation (28),[37] and good agreement between the measured and calculated weight loss was observed in their work. Based on the same assumption ($B = 0.07$ V), ΔW values were also calculated and are listed in Table 3 for comparison. Good agreement was obtained only at V = 1.70 m/s for the flow-through pipe channel and at V = 1.49 m/s for the annular flow channel. These flow velocities are close to that used previously by Syrett and Macdonald.[37] Large deviations occurred at other, particularly higher velocities. This is possibly due to a variation of B with velocity. Nevertheless, the maximum difference between the calculated and measured weight loss is a factor of slightly more than two. Syrett and Macdonald have also argued that, for the case in which the measured weight loss is substantially greater than the calculated value, the difference is probably due to the existence of very high corrosion rates at short exposure times (for example, see Figures 9 and 10).[37] Because the short exposure time region is the most difficult to define quantitatively in terms of the corrosion rate, the differences observed between the measured and calculated weight losses are of doubtful significance.

3.1.4 <u>Potentiodynamic Polarization Measurements</u>

Figures 37 through 39 show cathodic polarization curves at various Re numbers for the flow-through pipe channel, the annular flow channel, and the rotating cylinder system, respectively. Two plateaus (diffusion-limited currents) were observed in the cathodic region at each Re for each hydrodynamic system. One (I_{ℓ_1}), which occurs as a potential value closer to $E_{corrosion}$, was interpreted as being due to the reduction of oxygen diffusing through both the diffusion boundary layer (within the laminar sublayer) and a surface corrosion product film, while the other one (I_{ℓ_2}) was due to either the reduction of oxygen diffusing through the diffusion boundary layer only or to the reduction of oxygen diffusing through the diffusion boundary and an incompletely reduced surface film, depending on the value of the Re number. In all cases, the plateau current is controlled by mass transfer. As can be seen in these figures, the diffusion-limited current increases with increasing Re numbers.

Polarization curves were also determined potentiodynamically and potentiostatically for the annular flow channel at Re = 14,372 (Figure 38), in which no atacamite was found to accumulate on the alloy surface, and the sample was light red in color at the corrosion potential.[42] In this case, the potential was scanned at a rate of 25 mV/min from the corrosion potential into the cathodic region until the first plateau was reached. The potential was then kept constant for one hour. It was found that the current (I_{ℓ_1} at Re = 14,372) stayed constant and no change in color of the sample surface was observed during this potentiostatic period. However, as soon as the potential was decreased again to more negative values, the color of the sample surface changed from light red to red, and the current increased until the second plateau was reached. The potential was again kept constant for one hour, and a negligible increase in current with time was observed ($\sim$4%). It is then believed that for this case (Re = 14,372 for the annular flow channel), the first plateau (I_{ℓ_1}) corresponds to the diffusion-limited current for the reduction of oxygen diffusing through the diffusion boundary layer and a porous oxide (Cu_2O) film and is limited by the latter, whereas the second plateau (I_{ℓ_2}) corresponds, primarily, to the diffusion-limited current for the reduction of oxygen diffusing only through the diffusion boundary layer because the oxide film has been reduced.

Figures 40 through 42 show the variation of the diffusion-limited currents, I_{ℓ_1} and I_{ℓ_2}, as a function of exposure time at various Re numbers for the flow-through pipe channel, annular flow channel, and rotating cylinder system, respectively. The diffusion-limited currents, I_{ℓ_1}

and I_{ℓ_2}, were measured after 24 hours of exposure and were found to be nearly constant for high Re numbers, whereas, at low Re numbers, the currents decreased with increasing exposure time (increasing thickness of the surface film) and finally reached an almost constant value. The diffusion-limited currents, I_{ℓ_1} and I_{ℓ_2}, in the steady state were then averaged for each Re number and are referred to as i_{ℓ_1} and i_{ℓ_2}, respectively.

3.1.5 <u>Hydrodynamic Characteristics of the Systems</u>

Mass transfer studies of electrochemical reactions are normally carried out under mass transfer limited current conditions (e.g., oxygen reduction). When limiting conditions prevail, the mass transfer coefficient, K, can be expressed as:

$$K = \frac{I_\ell}{nFAC_b} \tag{29}$$

in which I_ℓ = Mass transfer limited current (amps),

  F = Faraday's constant,

  n = Number of electrons involved in the reaction (n = 4 for oxygen reduction),

  A = Area of the working electrode (cm^2), and

  C_b = Bulk concentration of the diffusion species ($O_2 \approx$ 8,000 ppb = 2.925 mole/cm^3 for air-saturated 1 M NaCl solution).

The Sherwood number (Sh) can be related to the mass transfer coefficient by:

$$Sh = \frac{KX}{D} \qquad\qquad (30)$$

in which D is the diffusivity of oxygen in 1 M NaCl solution, and X is:

- the inside diameter of the specimen for the flow-through pipe channel, or

- the effective diameter (i.e., inside diameter of the external acrylic tubing minus the outside diameter of the specimen) for the annular flow channel, or

- the outside diameter of the specimen for the rotating cylinder system.

Since the viscosity of a 1 M NaCl solution does not differ significantly from that of pure water, it is reasonable to assume that the diffusivity of oxygen in a 1 M NaCl solution is the same as that for oxygen in pure water. In the present study, D is taken to be $2.3 \cdot 10^{-5} cm^2/s$.

Values for the flow velocities (or the peripheral velocities) and the corresponding Re numbers, as well as the mass transfer coefficients, Sh numbers, and other experimental parameters calculated from the averaged steady state diffusion limiting currents, i_{ℓ_1} and i_{ℓ_2} , for each hydrodynamic system, are summarized in Tables 4 through 6.

3.1.6   Comparison with Published Data

3.1.6.1 The Flow-Through Pipe Channel

The hydrodynamic characteristics of the flow-through pipe channel are compared with those previously reported by Harriot and Hamilton,[13]

$$Sh = 0.0096 Re^{0.913} Sc^{0.346} \qquad (31)$$

by plotting $\log Sh_2 \cdot Sc^{-0.346}$ calculated from I_{t_2} against $\log Re$ (Figure 43). Good agreement is obtained in the high Re numbers region; however, in the low Re numbers region, data points fall below the established correlations with the deviation increasing with decreasing Re number. This is believed to be caused by the existence of a surface film, $Cu_2(OH)_3Cl$ (atacamite), formed at low Re numbers at the alloy/solution interface, which could resist the transport of oxygen to the metal. The deviation is more pronounced at low Re numbers, in which the surface film is thicker and results in a better barrier to oxygen transport.

It is well-known that for the Cu-Ni alloys, the corrosion resistance at low Re numbers depends on the formation of a protective barrier layer of Cu_2O,[38-40] upon which exists a film of atacamite that, in turn, protects the underlying film of Cu_2O.[42] It was also found by Popplewell that the atacamite layer $(Cu_2[OH]_3Cl)$ had deposited from the solution since its formation could be prevented by continuous filtering.[42] In this work, no atacamite was found at the high Re numbers region. The critical number, $Re_{critical}$, beyond which no deposition of atacamite occurred, and thus good agreement between the experimental data and that obtained by Harriot and Hamilton is observed, can be estimated by drawing a line to pass through the data points lying below the Harriot-Hamilton line in Figure 43 to intercept the Harriot-Hamilton line. The Re number was found to be approximately 10,000, which corresponds roughly to that for transition from laminar to turbulent flow.

3.1.6.2 *The Annular Flow Channel*

Ross and Wragg derived a solution for the mass transfer coefficient for annular flow under turbulent conditions:[18]

$$K = 0.807 D \left(\frac{B'}{DL} \right)^{\frac{1}{3}} \qquad (32)$$

in which L is the electrode length (= 1.27 cm) and B′ is the velocity gradient at the wall. In turbulent flow, B′ can be represented by:

$$B' = \frac{fV_\ell^2}{2v} \qquad (33)$$

in which V_ℓ is the average linear flow velocity, f is the friction factor, and v is the kinematic viscosity.

For the annular flow channel, it is necessary to use the appropriate equation for the friction factor, and this involves making the distinction between the friction factors at the inner and outer walls of the annulus, f_1 and f_2, respectively. This problem was discussed by Knudsen, who presented a method by which the friction factors at both walls of the annulus may be determined for different values of Re numbers and the ratio of inner and outer annulus radii, r_1/r_2:[43]

$$f_1 = 0.046\,Re^{-0.2}\left[\frac{1-\alpha}{1-\lambda^2}\right]^{0.2}\left[\frac{\lambda^2-\alpha^2}{\alpha(1-\lambda^2)}\right] \qquad (34)$$

$$f_2 = 0.046\,Re^{-0.2}\left[\frac{1-\alpha}{1-\lambda^2}\right]^{0.2} \qquad (35)$$

in which $\alpha = r_1/r_2$ and $\lambda = [(\alpha^2-1)/\ln(r^2)]^{0.5}$

Insertion of the appropriate values for α and λ ($\alpha = 0.75$, $\lambda = 0.872$), and substituting Equation (34) into (33), and then from Equation (32), with $Sh = KX/D$, we find:

$$Sh = 0.2373\,Re^{0.6}\,Sc^{\frac{1}{3}}\left(\frac{X}{L}\right)^{\frac{1}{3}} \qquad (36)$$

The experimentally derived characteristics of the annular flow channel were then compared with those predicted by Equation (36) by plotting $\log Sh_2 \cdot Sc^{-1/3}(X/L)^{-1/3}$, calculated from i_{ℓ_2} , vs $\log Re$ (Figure 44). Good agreement is obtained beyond a critical Re number ($\sim 10,000$) with deviations in the region of lower Re numbers. This deviation at lower Re numbers is also believed to be due to the existence of a surface film (film thickness increases with decreasing Re number), which impedes the transport of oxygen to the metal.

3.1.6.3 Rotating Cylinder System

The experimentally derived mass transfer rate is found to fit best with that established by Eisenberg et al. ($Sh = 0.0791\,Re^{0.7}Sc^{0.356}$) (Figure 45). Good agreement is also observed at high Re numbers ($Re_{critical} \approx 10,000$) with deviations occurring at the lowest Re number ($Re = 5,610$) studied.

The variation of log $Sh_2 \cdot Sc^{-0.356}$ with exposure time at Re = 5,610 is also included in the figure. At this Re number, good agreement is observed for a short exposure time (28.5 h). However, increasingly large deviations occur on longer exposure due to the increasing thickness of the surface film that resists the transport of oxygen to the surface.

3.1.7   Corrosion Rate/Mass Transfer Correlations

Figures 46 and 47 show the plots of corrosion rate (as $1/R_pA$), for all the hydrodynamic and geometrical systems, as a function of Sh_1 (calculated from i_{ℓ_1}) and Sh_2 (calculated from i_{ℓ_2}), respectively. Different correlations are obtained for different geometries. In addition, for each hydrodynamic system, the corrosion rate increases linearly with increasing Sh_1 and Sh_2 on a log-log scale. This reveals that the rate of corrosion is controlled by the rate of mass transfer. The corrosion rate is then plotted against Sh_h calculated from the mass transfer rate equations (13, 18, 26) previously established. It is also found that different curves are obtained for different geometries (Figure 48).

Figure 49 shows the variation of corrosion rate (as $1/R_pA$) as a function of mass transfer coefficients K_1, K_2, and K_h calculated from i_{ℓ_1}, i_{ℓ_2}, and the previously reported mass transfer rate equations (13, 18, 26), respectively. It can be seen from this figure that the corrosion rate increases linearly with increasing K_1 and K_2 on a log-log scale and is independent of the geometry involved. It is also found that above a critical value of K_h, K_{h_1} ($\approx 6 \cdot 10^{-3}$ cm/s), the corrosion rate measured in the flow-through pipe channel and the rotating cylinder system increases linearly with increasing K_h and is again independent of the geometry involved; and above another critical value of K_h, K_{h_1} ($\approx 1.5 \cdot 10^{-2}$ cm/s), the corrosion rate measured in the flow-through pipe channel and the annular flow channel (possibly also the rotating cylinder system) increases linearly with K_h and is also geometry independent. The Re numbers corresponding to the critical values for K_{h_1} and K_{h_2} are all approximately 10,000. This indicates that above a critical Re number ($\sim$10,000), the corrosion rate is independent of the geometries involved based on K_h. The deviations at lower Re numbers are probably due to the presence of a corrosion product film on the surface.

3.1.8 Transfer of Corrosion Rate from One System to Another

From the results shown in Figure 49, transfer of the corrosion rate measured in one system to another can be achieved based on either K_1 or K_h, and this implies that the corrosion rates in the flow-through pipe channel and the annular flow channel can be determined from the corrosion rates measured in the simple and low-cost rotating cylinder system.

Considering two mass transfer coefficients in series (Equation [37]), and with the surface film providing the greater resistance, the overall mass transfer coefficient K_1 may then be related to K_h by:

$$\frac{1}{K_1} = \frac{1}{K_h} + \frac{1}{K_f} \tag{37}$$

in which K_1 = overall mass transfer coefficient,

K_h = mass transfer coefficient for the diffusion boundary layer, and

K_f = mass transfer coefficient in the surface film.

Values for the mass transfer coefficient, K_f, were obtained from K_1 (calculated from experimental data) and K_h (calculated from previously published data) through Equation (37) and were then plotted against Re number on a log-log scale (Figure 50). The relationships obtained by linear regression analysis may be expressed as:

$$K_f = 8.36 \cdot 10^{-10}Re^{1.62} \text{ for the flow-through pipe channel,} \tag{38}$$

$$K_f = 2.37 \cdot 10^{-10}Re^{1.62} \text{ for the annular flow channel, and} \tag{39}$$

$$K_f = 1.30 \cdot 10^{-10}Re^{1.62} \text{ for the rotating cylinder} \tag{40}$$

by assuming that K_f has the same dependence on Re for these three geometries and by taking the average values of the slope as the exponent of Re. The dependencies of the mass transfer coefficients (K_h, K_1, and K_f) on the hydrodynamic parameters are summarized as follows:

- The Flow-Through Pipe Channel

$$Sh_h = 0.096Re^{0.913}Sc^{0.33} \tag{41}$$

$$K_h = 0.096Re^{0.913}Sc^{0.33}(D/X) \tag{42}$$

$$K_f = 8.36 \cdot 10^{-10}Re^{1.62} \tag{43}$$

$$K_1 = (1/K_h + 1/K_f)^{-1} = [1.04 \cdot 10^{-2}Re^{-0.913}Sc^{-0.33}(X/D) + 1.29 \cdot 10^{10}Re^{-1.62}]^{-1} \tag{44}$$

- The Annular Flow Channel

$$Sh_h = 0.2373Re^{0.6}Sc^{0.33}(X/L)^{0.33} \tag{45}$$

$$K_h = 0.2373Re^{0.6}Sc^{0.33}(X/L)^{0.33}(D/X) \tag{46}$$

$$K_f = 2.37 \cdot 10^{-10}Re^{1.62} \tag{47}$$

$$K_1 = (1/K_h + 1/3K_f)^{-1} = [4.21Re^{-0.6}Sc^{-0.33}(X/L)^{-0.33}(D/X)^{-1} + 3.23 \cdot 10^{9}Re^{-1.62}]^{-1} \tag{48}$$

- **The Rotating Cylinder System**

$$Sh_h = 0.0791 Re^{0.7} Sc^{0.356} \tag{49}$$

$$K_h = 0.0791 Re^{0.7} Sc^{0.356}(D/X) \tag{50}$$

$$K_f = 1.30 \cdot 10^{-10} Re^{1.62} \tag{51}$$

$$K_1 = (1/K_h + 1/K_f)^{-1} = [12.64 Re^{-0.7} Sc^{-0.356}(X/D) + 6.45 \cdot 10^{9} Re^{-1.62}]^{-1} \tag{52}$$

3.1.8.1 Transfer of Corrosion Rate Based on K_1

Since the corrosion rate measured in all the hydrodynamic systems studied is independent of the geometry involved, when plotted as a function of the overall mass transfer coefficient K_1 on a log-log scale, transfer of the corrosion rate from one system (rotating cylinder) to another (the flow-through pipe channel or the annular flow channel) can thus be made based on K_1. This can be achieved by inserting the appropriate values of Re, Sc, X, L, and D into Equation (48) or Equation (52) to calculate the value of K_1 for the system of interest. From the plot of corrosion rate vs K_1 obtained for the rotating cylinder system, the corrosion rate for the system of interest, either the flow-through pipe channel or the annular flow channel, can then be determined. However, the relation between K_f and Re for the system of interest must be known.

3.1.8.2 Transfer of Corrosion Rate Based on K_h

Because the corrosion rate measured in all the geometrical systems studied is independent of the geometry involved (on a plot of $\log[1/R_pA]$ vs $\log K_h$ in the region for which $K_h > K_{h,2}$ in which no atacamite deposition occurred), transfer of the corrosion rate from one system to another can thus also be made based on K_h. This can be achieved by inserting the appropriate values of Re, Sc, D, X, and L into Equation (42) or Equation (46) to calculate the value of K_h for the system of interest. From the plot of corrosion rate vs K_h obtained in the rotating cylinder system, the corrosion rate for the system of interest can then be determined.

3.2 THE EFFECT OF DISSOLVED OXYGEN CONCENTRATION ON CORROSION RATE

The effect of dissolved oxygen concentration on the corrosion rate of 90:10 Cu-Ni alloy in a flowing 1 M NaCl solution at pH = 5, T = 25°C (77°F) has been studied at fixed flow velocity for the flow-through pipe channel (V_t = 1.70 m/s) and the annular flow channel(V_t = 1.49 m/s).

3.2.1   Variation of Corrosion Potential With Time

Figures 51 and 52 show the variation of the corrosion potential, $E_{corrosion}$, as a function of exposure time at various dissolved oxygen concentrations for the flow-through pipe channel and the annular flow channel, respectively.

Under air-saturated conditions (oxygen content ≈ 8,000 ppb), for both channels, the corrosion potential increased in the more noble direction within the first few hours of exposure and then decreased and stabilized. However, for all other oxygen contents studied, the corrosion potential increased with increasing exposure time from the beginning and finally reached a nearly constant value. The values of the corrosion potential after the first 72 hours of exposure were averaged for each oxygen concentration and then plotted as a function of oxygen concentration (Figure 53). It can be seen that the corrosion potential first increases but then decreases with increasing oxygen concentration. This can be explained as follows: At low oxygen concentrations, increasing the dissolved oxygen concentration can increase the availability of the oxygen at the interface, which will polarize the corrosion reaction to a more noble value. The decrease of $E_{corrosion}$ at higher oxygen concentrations shown in Figure 53 is believed to be due to the formation of a more protective surface film, which offers better resistance to the transport of oxygen to the interface.

3.2.2   Corrosion Rate Measurements

3.2.2.1 *Small Amplitude Cyclic Voltammetry (SACV)*

Figures 54 through 57 and Figures 58 through 61 show plots of corrosion rate as a function of exposure time and oxygen concentration for the flow-through pipe channel ($V_{,}$ = 1.70 m/s) and the annular flow channel ($V_{,}$ = 1.49 m/s), respectively. It is clear from Figures 57 and 61 (air-saturated conditions) that, for both test channels, the corrosion rate increased in the first few hours of exposure and then decreased to a relatively constant value. Although this pattern was not always found for all other oxygen concentrations studied, the corrosion rate tended towards a nearly constant value for long exposure times.

3.2.2.2 *Alternating Current (AC) Impedance*

AC impedance measurements were performed at various time intervals during each run. Figures 62 through 65 and Figures 66 through 69 show the complex plane impedance diagrams at different dissolved oxygen contents and exposure times for the flow-through pipe channel and the annular flow channel, respectively.

Inductance loops in the low-frequency region of the complex impedance diagrams were always observed at all oxygen concentrations studied. However, no clearly defined intercept was observed in the low-frequency region under air-saturated conditions for the annular flow channel (Figure 69), and no reasonable estimation of R_s (high-frequency intercept) could be made under air-saturated conditions for the flow-through pipe channel (Figure 65). In these cases, the AC impedance data had to be discarded.

For all other dissolved oxygen concentrations studied, the curves in the high-frequency region were extrapolated in order to obtain R_s. The maximum error in estimated R_s was 2 ohms. The value of R_p was then obtained by subtracting the high-frequency intercept (R_s) from the low-frequency intercept ($R_s + R_p$).

Values for $1/R_p$ obtained from AC impedance measurements for each channel were plotted as functions of exposure time. They are included in Figures 54 through 61 for comparison with those obtained from SACV. Good agreement between these two techniques was observed. For each oxygen concentration, the values of $1/R_p$ at the steady state obtained with SACV were then averaged and plotted against oxygen concentration (Figure 70). The corrosion rate first increases but then decreases with increasing oxygen concentration. At low dissolved oxygen concentrations, for the velocity studied ($V_t = 1.70$ m/s for the flow-through pipe channel and $V_t = 1.49$ m/s for the annular flow channel), the effect of increasing dissolved oxygen concentration is to increase the interfacial concentration of the reactant (oxygen), thereby leading to a higher corrosion rate. The decrease in the corrosion rate observed at high oxygen concentrations shown in Figure 70 can be attributed to the formation of a more protective passive film, which reduces the availability of oxygen to the reacting interface.

3.2.2.3 Weight Loss Measurements

Table 7 summarizes the weight losses calculated from the values of $1/R_p$ obtained from SACV by Equation (28) (assuming $B = 0.07$ volt) for the flow-through pipe channel and annular flow channel at each dissolved oxygen concentration studied. They are in reasonable accord with the directly measured weight losses.

3.3 THE EFFECT OF PH ON THE CORROSION POTENTIAL AND CORROSION RATE

The effect of pH on the corrosion potential and corrosion rate of 90:10 Cu-Ni alloy in a flowing 1 M NaCl solution at room temperature ($25°C$ [$77°F$]) and at a constant low velocity ($V_t = 1.70$ m/s for the flow-through pipe channel and $V_t = 1.49$ m/s for the annular flow channel) has been studied for the flow-through pipe channel and the annular flow channel under air-saturated conditions at pH = 5, 6, 7, and 8.5.

3.3.1 Variation of Corrosion Potential with Time

Figure 71 shows the variation of corrosion potential, $E_{corrosion}$, as a function of exposure time at various pH values for the flow-through pipe channel ($V_t = 1.70$ m/s) and the annular flow channel ($V_t = 1.49$ m/s) under air-saturated conditions. For both channels, at pH = 5 and pH = 6, the corrosion potential increased in the more noble direction within the first four to five hours of exposure and then decreased and stabilized.

However, at pH = 7 the corrosion potential decreased from the very beginning of exposure and then became constant. For pH = 8.5, the values for the corrosion potential were found to be scattered for the initial period of exposure. Nevertheless, they tended towards a constant value for long exposure times.

The steady-state corrosion potentials measured in both channels were then plotted as a function of pH (Figure 72). For both channels, the corrosion potential first decreases with increasing pH, but then apparently tends to a constant value at pH values above 7.

3.3.2   Corrosion Rate Measurements

3.3.2.1 Small Amplitude Cyclic Voltammetry (SACV)

Figures 73 through 76 and Figures 77 through 80 show the variation of corrosion rate with exposure time at a fixed flow velocity and at various pH values under air-saturated conditions for the flow-through pipe channel and the annular flow channel, respectively. It is clear from these figures that for both test channels, and at both pH = 5 and pH = 6, the corrosion rate increased during the first few hours of exposure but subsequently decreased and finally became nearly constant. For pH = 7, the corrosion rate decreased gradually from the very beginning and finally stabilized after 90 hours of exposure. For pH = 8.5, scattering of the corrosion rate with respect to exposure time was observed in the initial exposure period; however, the corrosion rate became constant after long exposure times.

3.3.2.2 *Alternating Current (AC) Impedance*

Figures 81 through 84 and Figures 85 through 88 show the complex plane impedance diagrams at different pH values and exposure times for the flow-through pipe channel and the annular flow channel, respectively. No clearly defined intercept was found for pH = 5, 6, and 7, either in the low-frequency region or in the high-frequency region, and the AC impedance data obtained for these pH values had to be discarded. For pH = 8.5, the curves in the complex plane impedance diagrams were extrapolated in order to calculate R_p. Variation of $1/R_p$ obtained by the AC impedance technique with exposure time at pH = 8.5 was also included in Figures 76 and 80 for the flow-through pipe channel and the annular flow channel, respectively. As can be seen from these two figures, good agreement was observed between the SACV and AC impedance technique for the measurement of polarization resistance.

For each pH value, the corrosion rate (as $1/R_p$) at the steady state obtained in SACV was averaged and plotted against the pH of the solution (Figure 89). For both channels, the rate of corrosion first decreases with increasing pH but then tends towards a constant value at pH values above 6.

3.3.2.3 *Weight Loss Measurement*

The weight losses calculated from the values of $1/R_p$ obtained in SACV by Equation (28) (assuming B = 0.07 volt) for the flow-through pipe channel and the annular flow channel at each pH value are listed in Table 8. They are in reasonable accord with the directly measured weight losses also included in this table.

TABLE 3

Comparison Between Measured and Calculated Weight Loss Under Air-Saturated Conditions at Various Velocities

Linear Flow Velocity (V_ℓ) or Peripheral Velocity (U) and Reynolds number (Re)	Exposure Time (Hours)	A_ℓ (Coulombs /volt)	Weight Loss ΔW (g) Calculated	Weight Loss ΔW (g) Measured
THE FLOW-THROUGH PIPE CHANNEL				
V_ℓ = 0.31 m/s, Re = 4,624	94.0	5,164.2	0.09	0.043
V_ℓ = 0.65 m/s, Re = 9,695	148.5	22,804.4	0.41	0.294
V_ℓ = 1.70 m/s, Re = 25,355	168.0	47,663.6	0.86	0.784
V_ℓ = 2.75 m/s, Re = 41,016	130.2	48,453.5	0.87	1.025
V_ℓ = 4.37 m/s Re = 65,178	150.0	74,019.8	1.33	2.047
THE ANNULAR FLOW CHANNEL				
V_ℓ = 0.27 m/s, Re = 1,624	94.0	4,394.6	0.08	0.041
V_ℓ = 0.56 m/s, Re = 3,367	148.5	20,628.1	0.37	0.281
V_ℓ = 1.49 m/s, Re = 8,960	168.0	32,473.4	0.58	0.477
V_ℓ = 2.39 m/s, Re = 14,372	130.2	47,217.5	0.85	1.257
V_ℓ = 3.79 m/s, Re = 22,790	150.0	87,054.2	1.56	2.351
THE ROTATING CYLINDER SYSTEM				
U = 0.6 m/s, Re = 10,824	129.8	16,701.2	0.30	0.216
U = 1.0 m/s, Re = 18,040	104.2	22,591.5	0.41	0.108
U = 2.0 m/s, Re = 36,080	150.0	35,105.6	0.63	0.394

TABLE 4

Summary of Experimental Parameters
for the Flow-Through Pipe Channel

V_ℓ (m/s)		0.31	0.65	1.70	2.75	4.37
Re		4,624	9,695	25,355	41,016	65,178
i_{ℓ_1} (mA)		0.095	0.58	2.40	3.67	6.28
i_{ℓ_2} (mA)		0.23	2.12	8.07	12.39	21.85
$K_1 (\cdot 10^{-3} \text{cm/s})$		0.134	0.818	3.383	5.174	8.853
$K_2 (\cdot 10^{-3} \text{cm/s})$		0.324	2.989	1.138	1.747	3.080
$K_h (\cdot 10^{-3} \text{cm/s})$		2.659	5.224	12.529	19.451	29.677
$K_f (\cdot 10^{-3} \text{cm/s})$		0.141	0.970	4.634	7.048	12.617
Sh_1		9.18	56.02	231.66	354.27	606.24
Sh_2		22.19	204.68	779.01	1196.00	2109.20
Sh_h		182.06	357.71	857.94	1332.00	2032.20
$Sh_2 \cdot Sc^{-0.346}$		2.66	24.55	93.45	143.48	253.02
$1/R_p$ $(\cdot 10^{-2}$ ohms$^{-1})$	SACV	0.75	3.46	7.75	10.20	14.40
	AC Impedance	X	3.39	X	10.87	15.71
$1/R_p \cdot 1/A$ $(\cdot 10^{-2}$ ohms^{-1} cm$^{-2})$	SACV	0.12	0.55	1.23	1.62	2.29
	AC Impedance	X	0.54	X	1.73	2.50

X = data discarded

TABLE 5

Summary of Experimental Parameters
for the Annular Flow Channel

V_ℓ (m/s)		0.27	0.56	1.49	2.39	3.79
Re		1,624	3,367	8,960	14,372	22,790
i_{ℓ_1} (mA)		0.13	0.35	1.07	3.81	7.31
i_{ℓ_2} (mA)		0.45	1.28	3.58	12.88	25.20
K_1 ($\cdot 10^{-3}$cm/s)		0.146	0.408	1.247	4.440	8.518
K_2 ($\cdot 10^{-3}$cm/s)		0.524	1.492	4.172	15.008	29.364
K_h ($\cdot 10^{-3}$cm/s)		4.532	7.060	12.542	16.750	22.104
K_f ($\cdot 10^{-3}$cm/s)		0.151	0.433	1.384	6.041	13.858
Sh_1		4.02	11.26	34.42	122.57	235.17
Sh_2		14.48	41.18	115.17	414.35	810.70
Sh_h		125.1	194.91	346.28	462.44	610.26
$Sh_2\, Sc^{-1/3}\, (X/L)^{-1/3}$		2.37	6.74	18.84	67.79	132.63
$1/R_p$ ($\cdot 10^{-2}$ ohms^{-1})	SACV	0.87	1.85	5.25	10.1	16.90
	AC Impedance	X	2.84	X	10.20	16.30
$1/R_p A$ ($\cdot 10^{-2}$ ohms^{-1} cm^{-2})	SACV	0.12	0.24	0.69	1.33	2.22
	AC Impedance	X	0.37	X	1.34	2.14

X = data discarded

TABLE 6

Summary of Experimental Parameters for the Rotating Cylinder System

U (m/s)		0.60	1.00	2.00
Re		10,824	18,040	36,080
i_{ℓ_1} (mA)		0.25	1.15	1.83
i_{ℓ_2} (mA)		0.74	4.35	7.64
$K_1(\cdot 10^{-3} cm/s)$		0.291	1.340	2.132
$K_2(\cdot 10^{-3} cm/s)$		0.862	5.069	8.203
$K_h(\cdot 10^{-3} cm/s)$		5.710	8.149	132.63
$K_f(\cdot 10^{-3} cm/s)$		0.446	1.830	2.860
Sh_1		24.13	110.99	176.62
Sh_2		71.42	419.83	679.45
Sh_h		472.90	674.93	1,098.50
$Sh_2 \cdot Sc^{-0.356}$		8.06	47.37	76.66
$1/R_p$ ($\cdot 10^{-2}$ ohms)	SACV	2.28	4.25	6.38
	AC Impedance	2.00	4.58	9.87
$1/R_pA$ ($\cdot 10^{-2}$ ohms^{-1} cm^{-2})	SACV	0.30	0.56	0.84
	AC Impedance	0.26	0.60	1.30

TABLE 7

Comparison Between Measured and Calculated Weight Losses at Different Oxygen Concentrations at pH = 5

$[O_2]$ (ppb)	Exposure Time (Hours)	A_t (Coulomb/ Volt)	Weight Loss ΔW (g)	
			Calculated	Measured
THE FLOW-THROUGH PIPE CHANNEL ($V_t = 1.70$ m/s, Re = 25,355)				
8,000	168.0	47,663.6	0.86	0.785
200	125.5	58,334.5	1.05	0.819
60	120.0	23,470.3	0.42	0.219
40	129.0	26,245.1	0.47	0.132
THE ANNULAR FLOW CHANNEL ($V_t = 1.49$ m/s, Re = 8,960)				
8,000	168.0	32,473.4	0.58	0.477
200	125.5	69,545.9	1.25	0.931
60	120.0	13,196.5	0.24	0.220
40	129.8	22,387.6	0.40	0.137

TABLE 8

Comparison Between Measured and Calculated Weight Losses at Different pH Values Under Air-Saturated Conditions

pH	Exposure Time (Hours)	A_t (Coulomb/ Volt)	Weight Loss ΔW (g)	
			Calculated	Measured
THE FLOW-THROUGH PIPE CHANNEL $(V_t = 1.70$ m/s, Re $= 25,355)$				
5	168.0	47,663.6	0.86	0.785
6	172.3	4,714.4	0.08	0.101
7	197.5	3,989.7	0.07	0.023
8.5	147.0	2,673.1	0.05	0.020
THE ANNULAR FLOW CHANNEL $(V_t = 1.49$ m/s, Re $= 8,960)$				
5	168.0	32,473.4	0.58	0.477
6	172.3	3,442.8	0.06	0.022
7	197.5	4,227.1	0.08	0.024
8.5	147.0	3,501.1	0.06	0.039

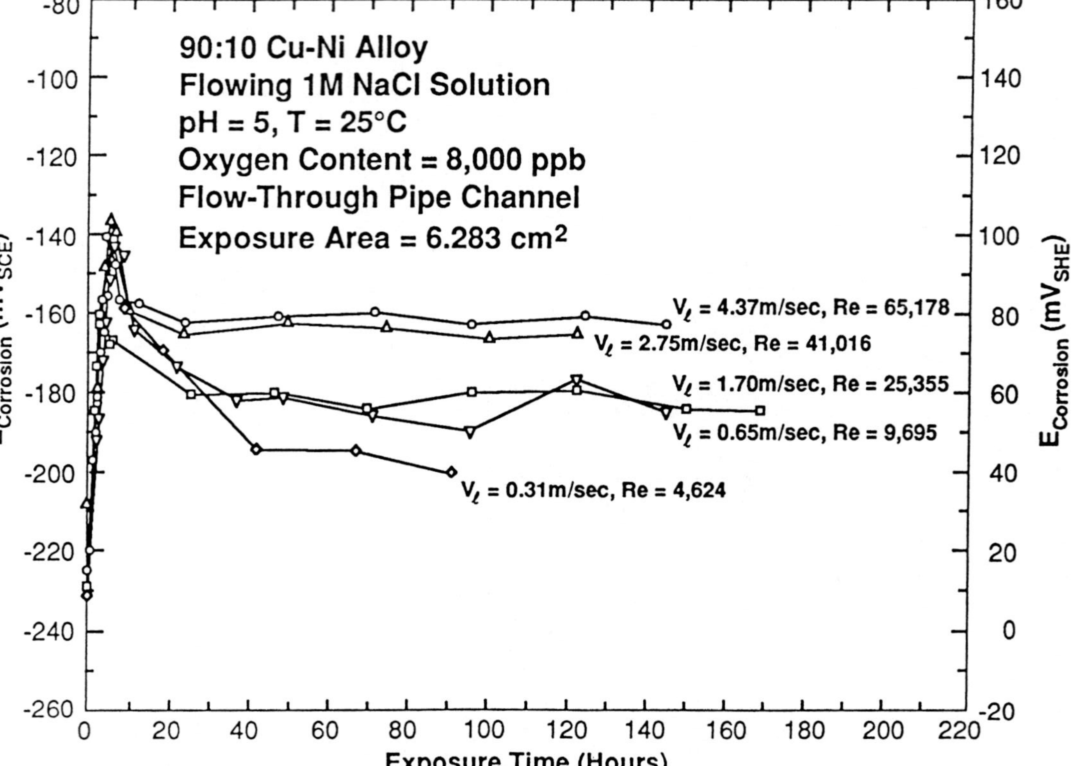

Figure 5   Variation of Corrosion Potential with Exposure Time at Various Reynolds Numbers Under Air-Saturated Conditions, at pH = 5, T = 25°C (77°F), for the Flow-Through Pipe Channel

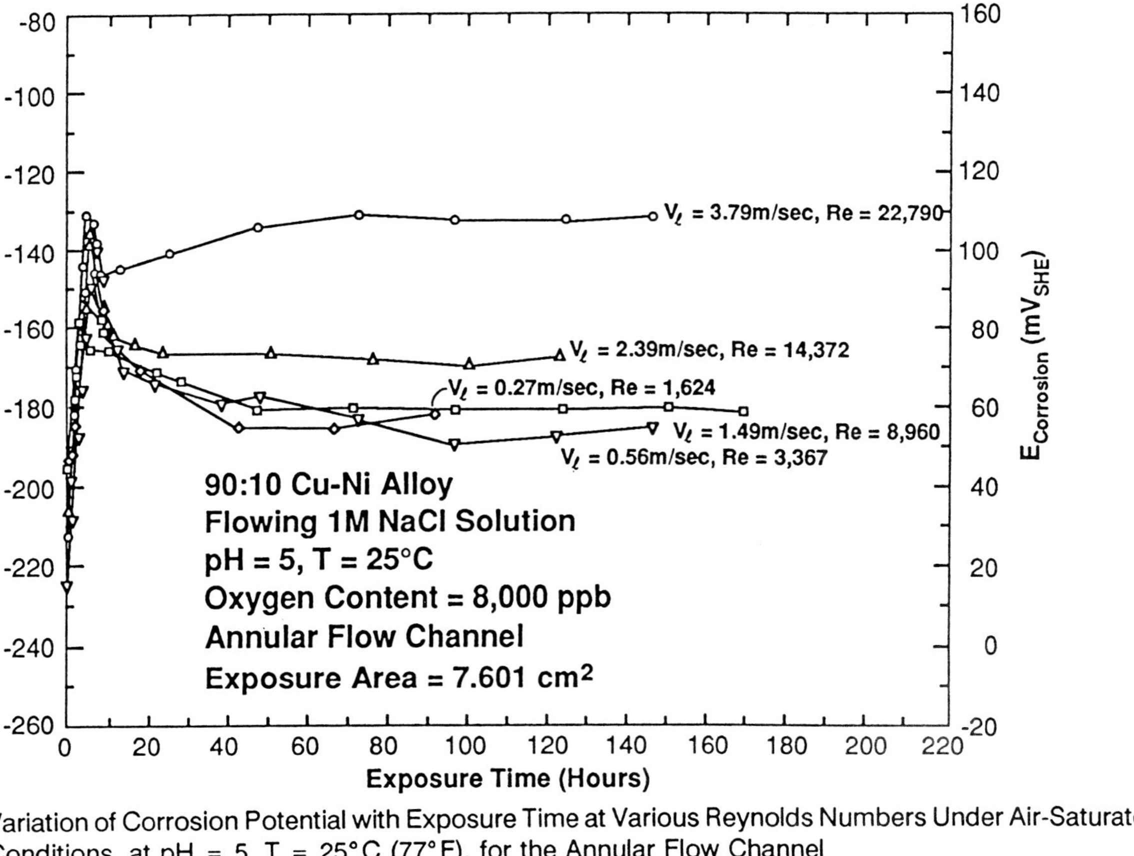

Figure 6   Variation of Corrosion Potential with Exposure Time at Various Reynolds Numbers Under Air-Saturated Conditions, at pH = 5, T = 25°C (77°F), for the Annular Flow Channel

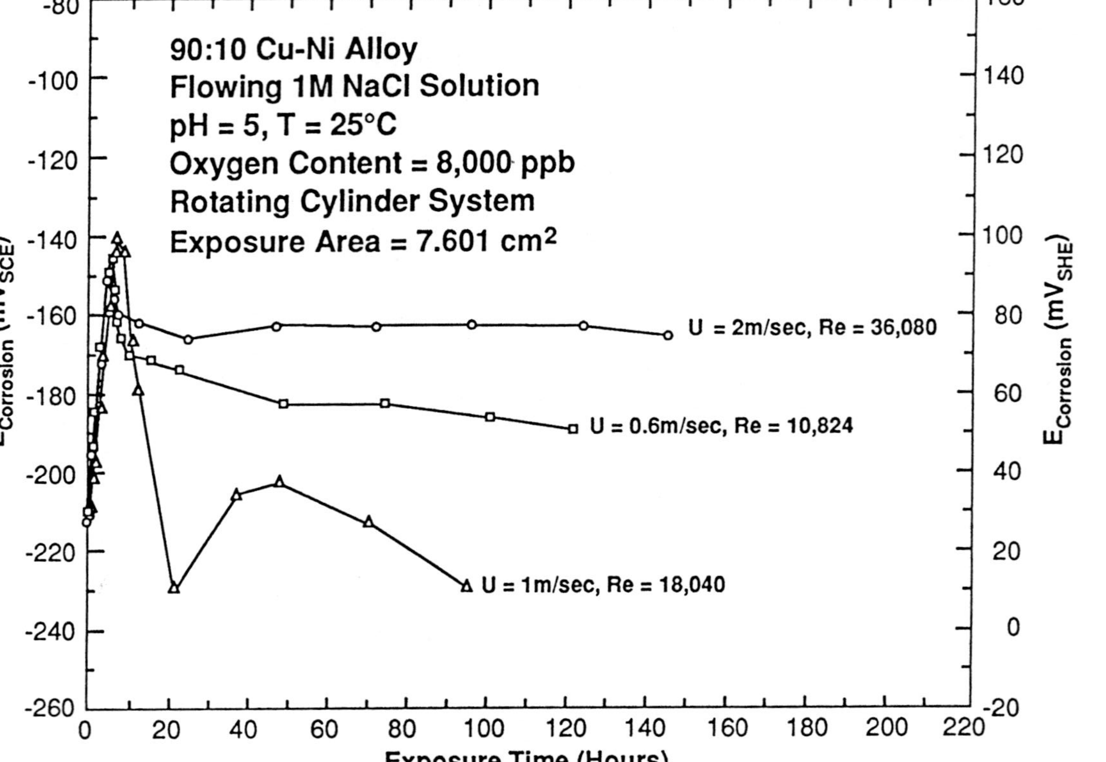

Figure 7   Variation of Corrosion Potential with Exposure Time at Various Reynolds Numbers Under Air-Saturated Conditions, at pH = 5, T = 25°C (77°F), for the Rotating Cylinder System

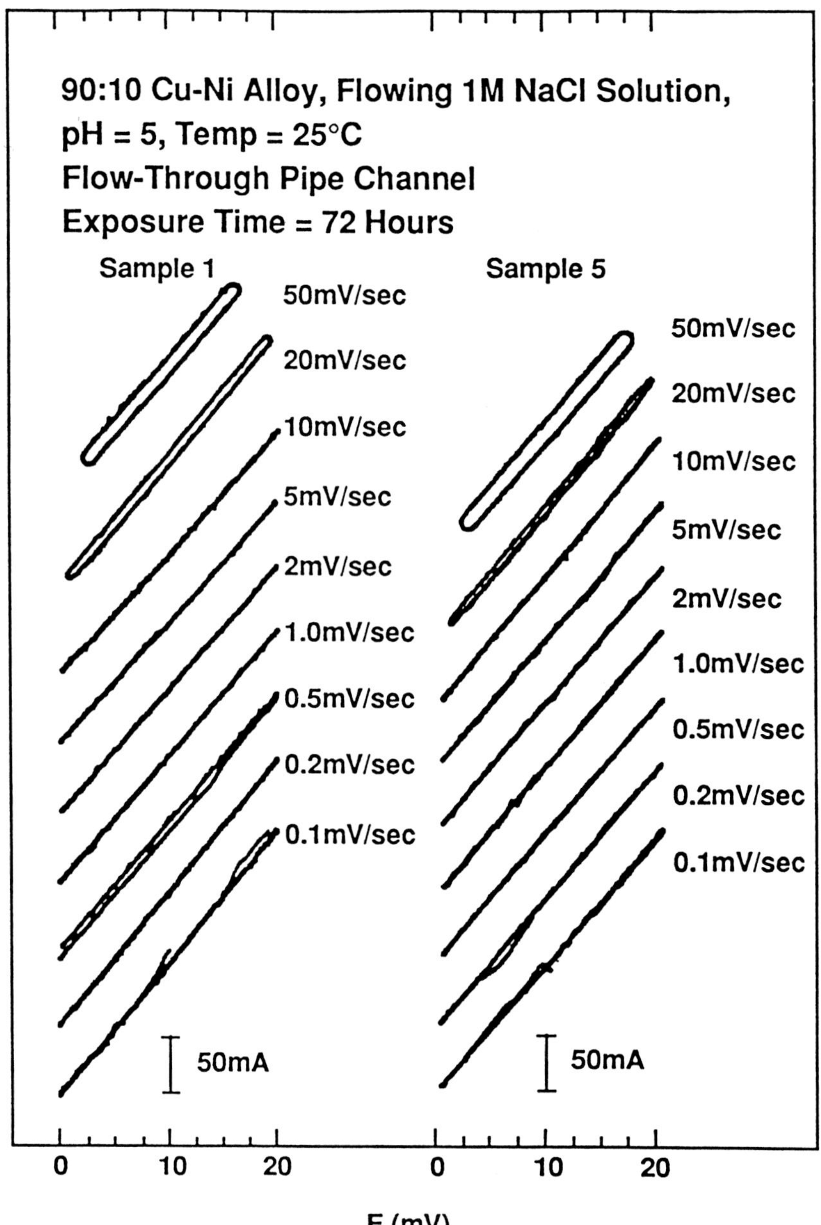

Figure 8   Typical Small Amplitude Cyclic Voltammogram for 90:10 Cu-Ni Alloy in Flowing Air-Saturated 1 M NaCl Solution at pH = 5, T = 25°C (77°F)

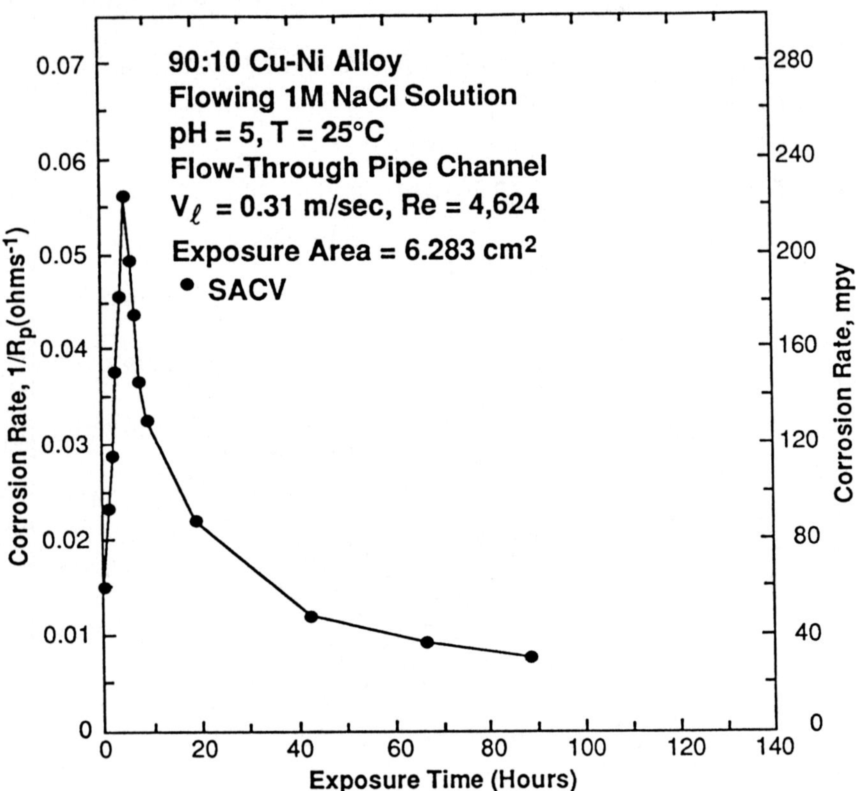

Figure 9   Variation of Corrosion Rate (as $1/R_p$) with Exposure Time Under Air-Saturated Conditions, at pH = 5, T = 25°C (77°F), for the Flow-Through Pipe Channel at Re = 4,624

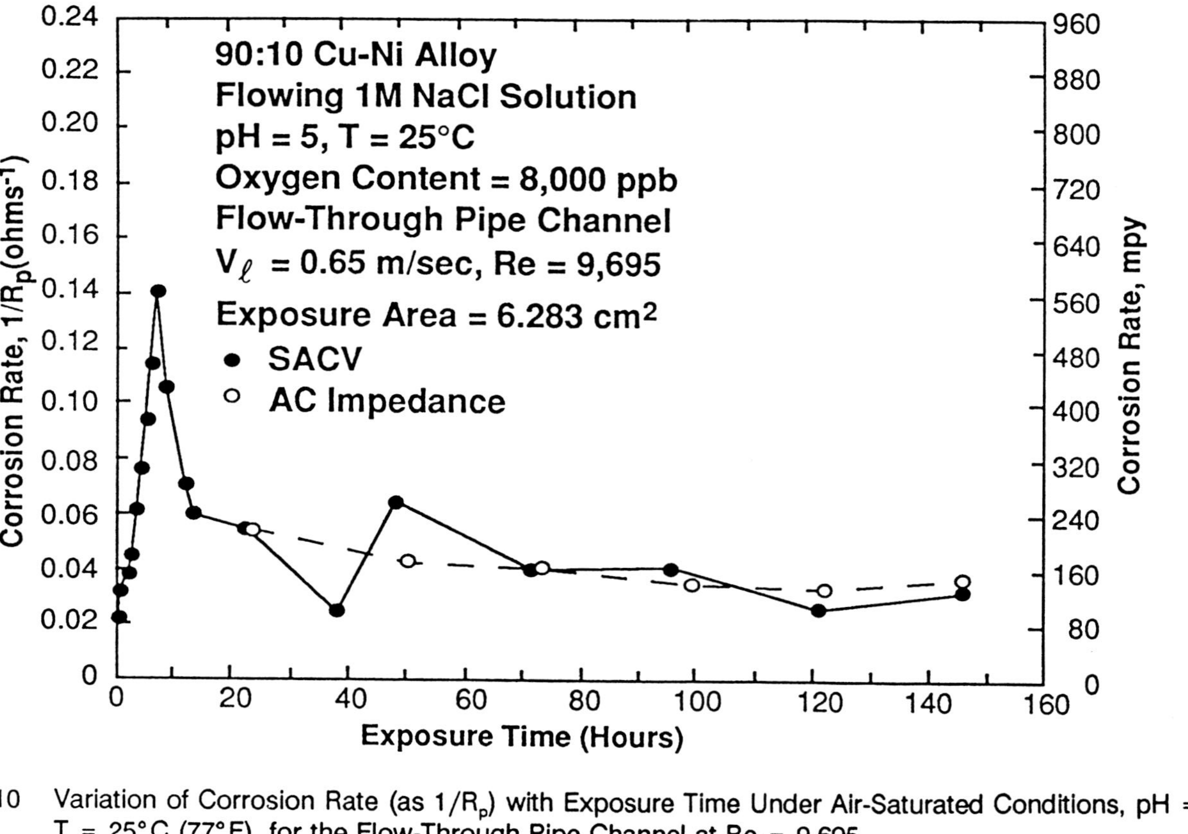

Figure 10   Variation of Corrosion Rate (as $1/R_p$) with Exposure Time Under Air-Saturated Conditions, pH = 5, T = 25°C (77°F), for the Flow-Through Pipe Channel at Re = 9,695

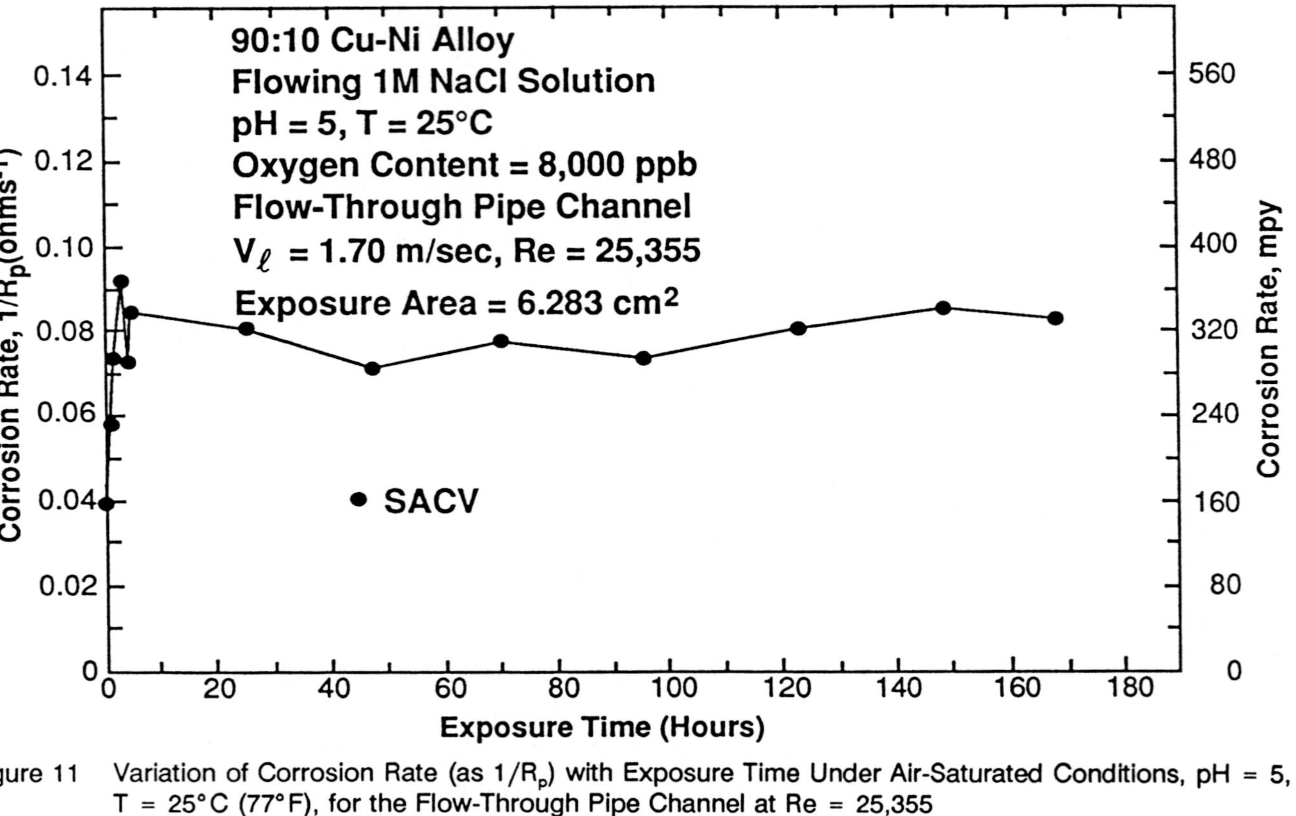

Figure 11   Variation of Corrosion Rate (as 1/R_p) with Exposure Time Under Air-Saturated Conditions, pH = 5, T = 25°C (77°F), for the Flow-Through Pipe Channel at Re = 25,355

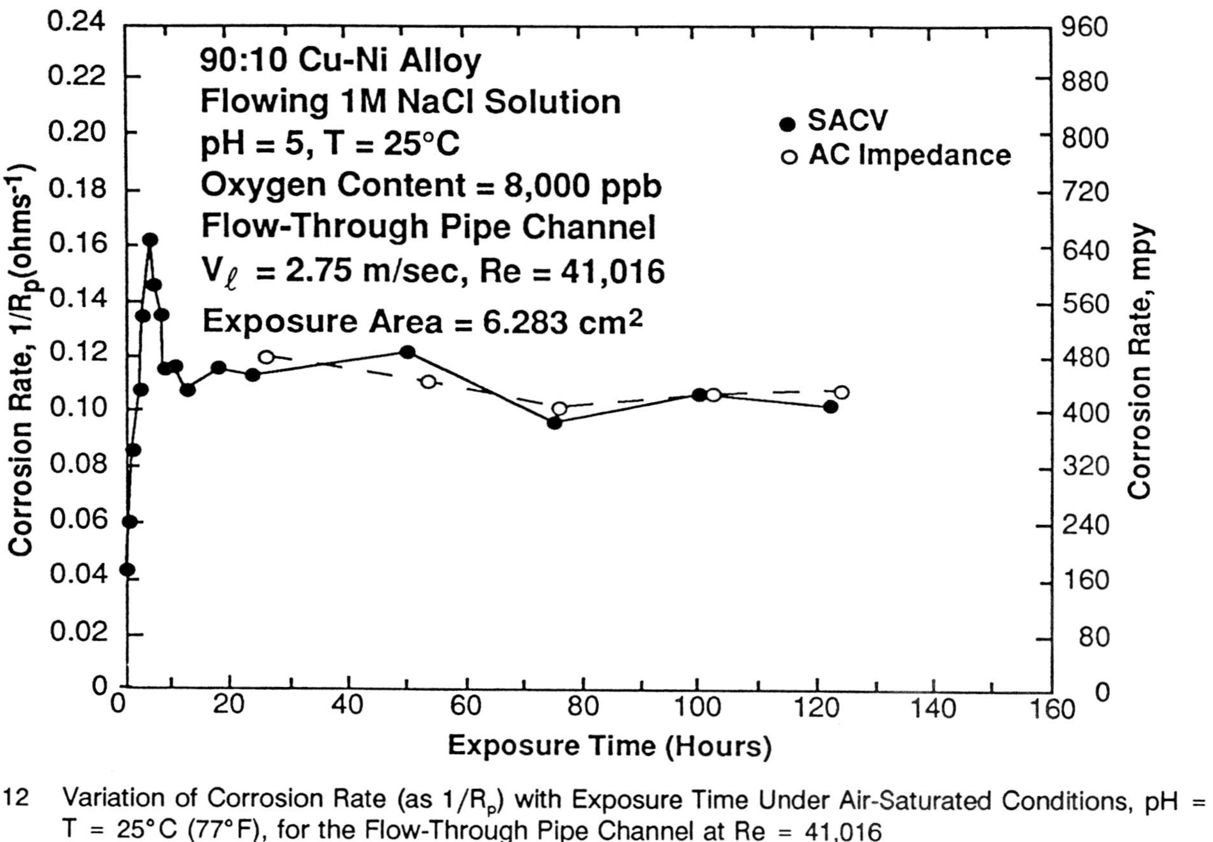

Figure 12   Variation of Corrosion Rate (as 1/R$_p$) with Exposure Time Under Air-Saturated Conditions, pH = 5, T = 25°C (77°F), for the Flow-Through Pipe Channel at Re = 41,016

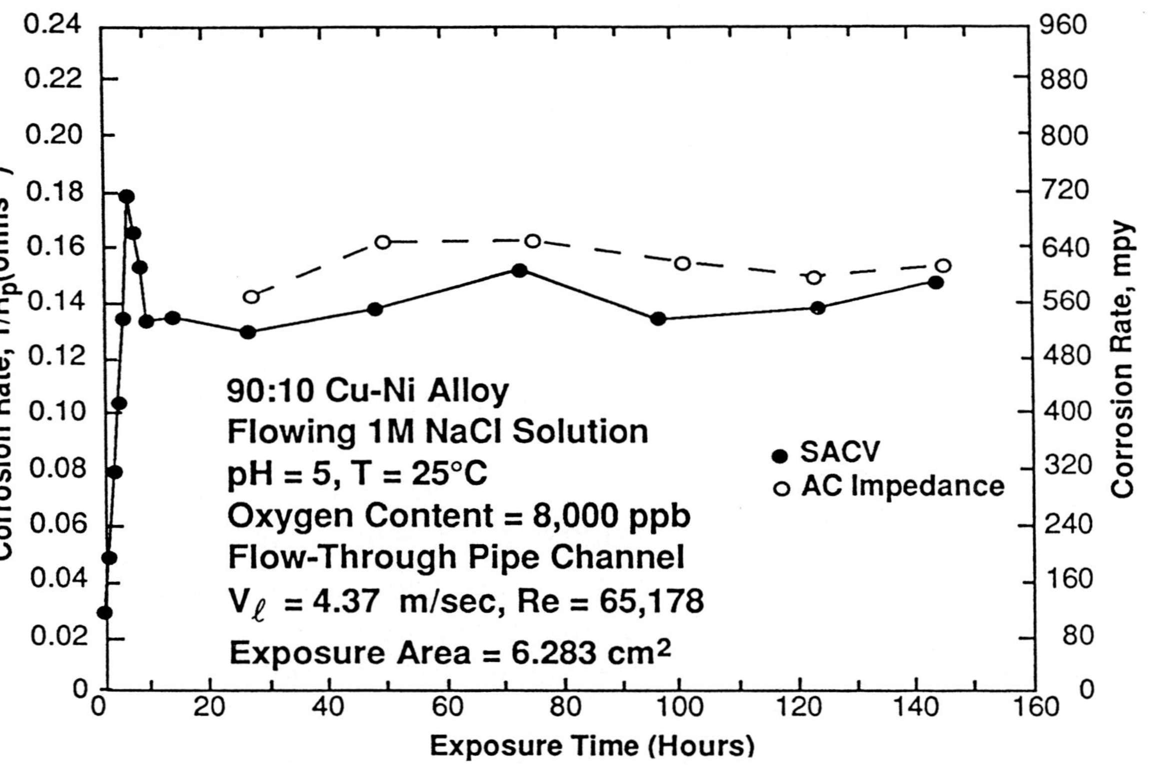

Figure 13   Variation of Corrosion Rate (as 1/R$_p$) with Exposure Time Under Air-Saturated Conditions, pH = 5, T = 25°C (77°F), for the Flow-Through Pipe Channel at Re = 65,178

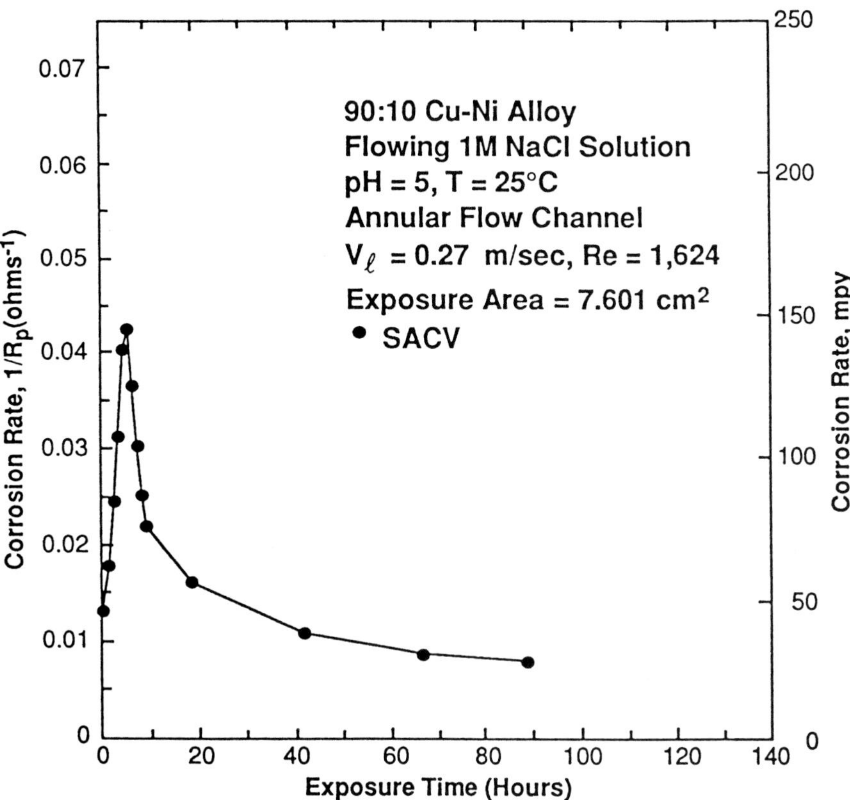

Figure 14   Variation of Corrosion Rate (as 1/R_p) with Exposure Time Under Air-Saturated Conditions, pH = 5, T = 25°C (77°F), for the Annular Flow Channel at Re = 1,624

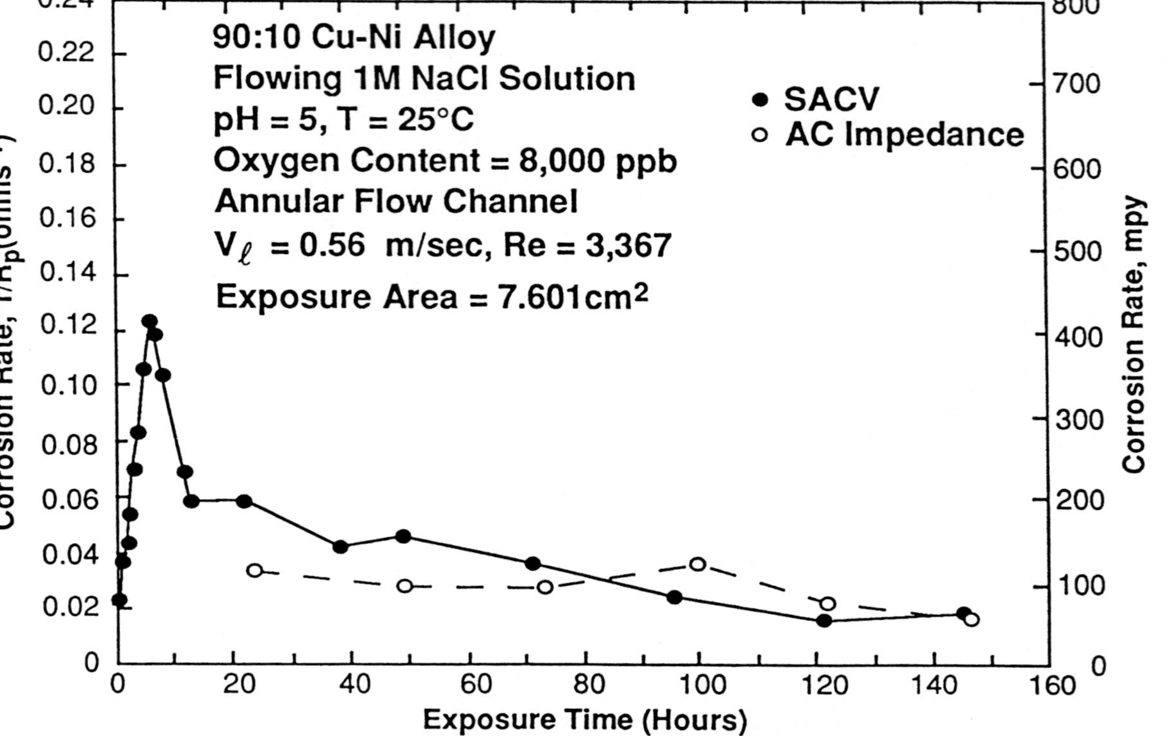

Figure 15   Variation of Corrosion Rate (as $1/R_p$) with Exposure Time Under Air-Saturated Conditions, pH = 5, T = 25°C (77°F), for the Annular Flow Channel at Re = 3,367

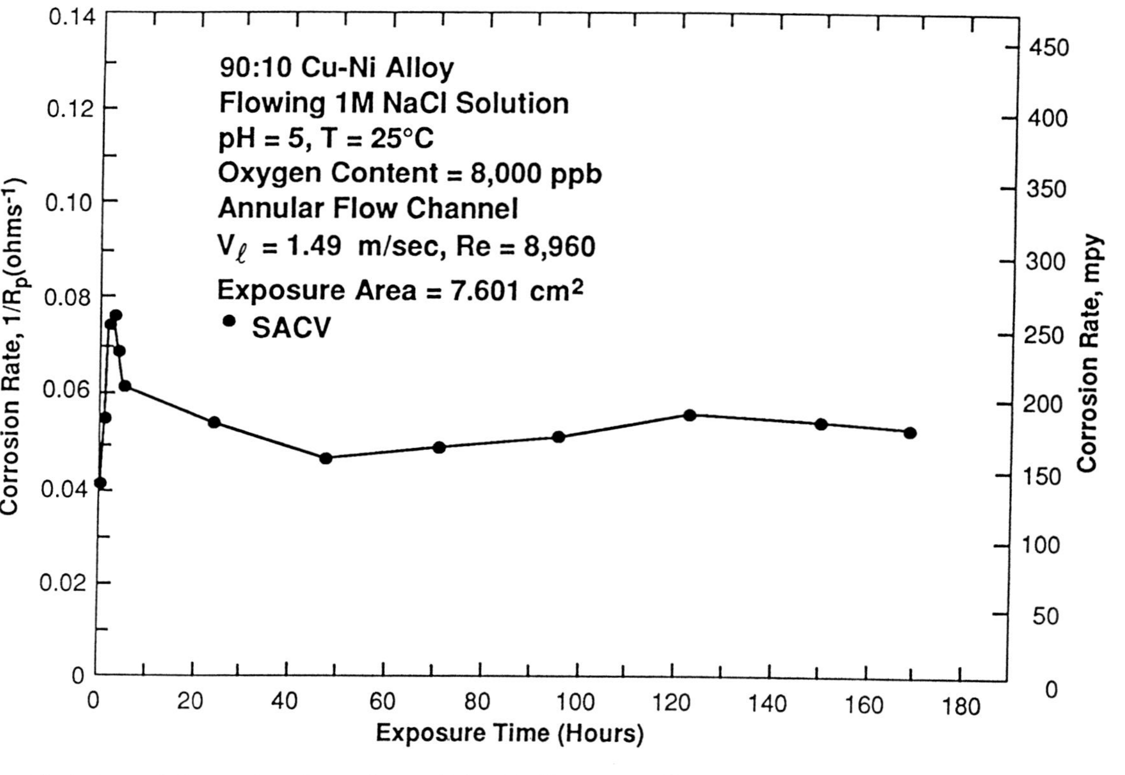

Figure 16   Variation of Corrosion Rate (as $1/R_p$) with Exposure Time Under Air-Saturated Conditions, pH = 5, T = 25°C (77°F), for the Annular Flow Channel at Re = 8,960

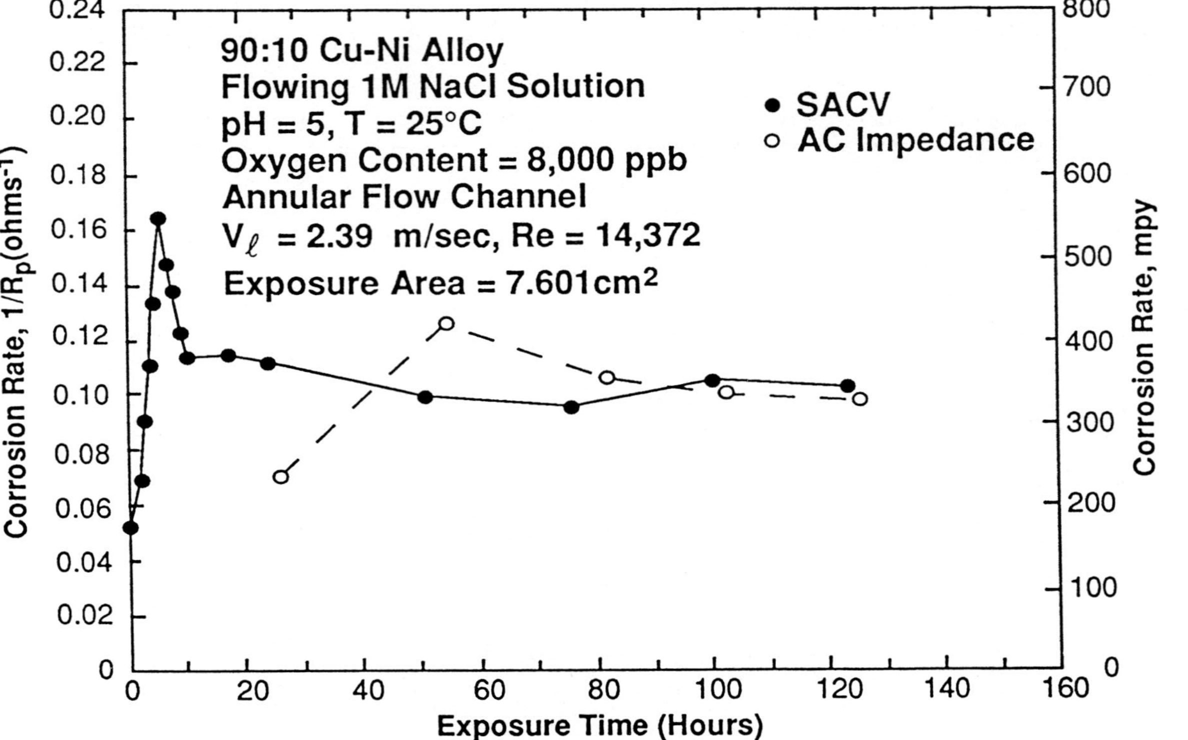

Figure 17   Variation of Corrosion Rate (as 1/R_p) with Exposure Time Under Air-Saturated Conditions, pH = 5, T = 25°C (77°F), for the Annular Flow Channel at Re = 14,372

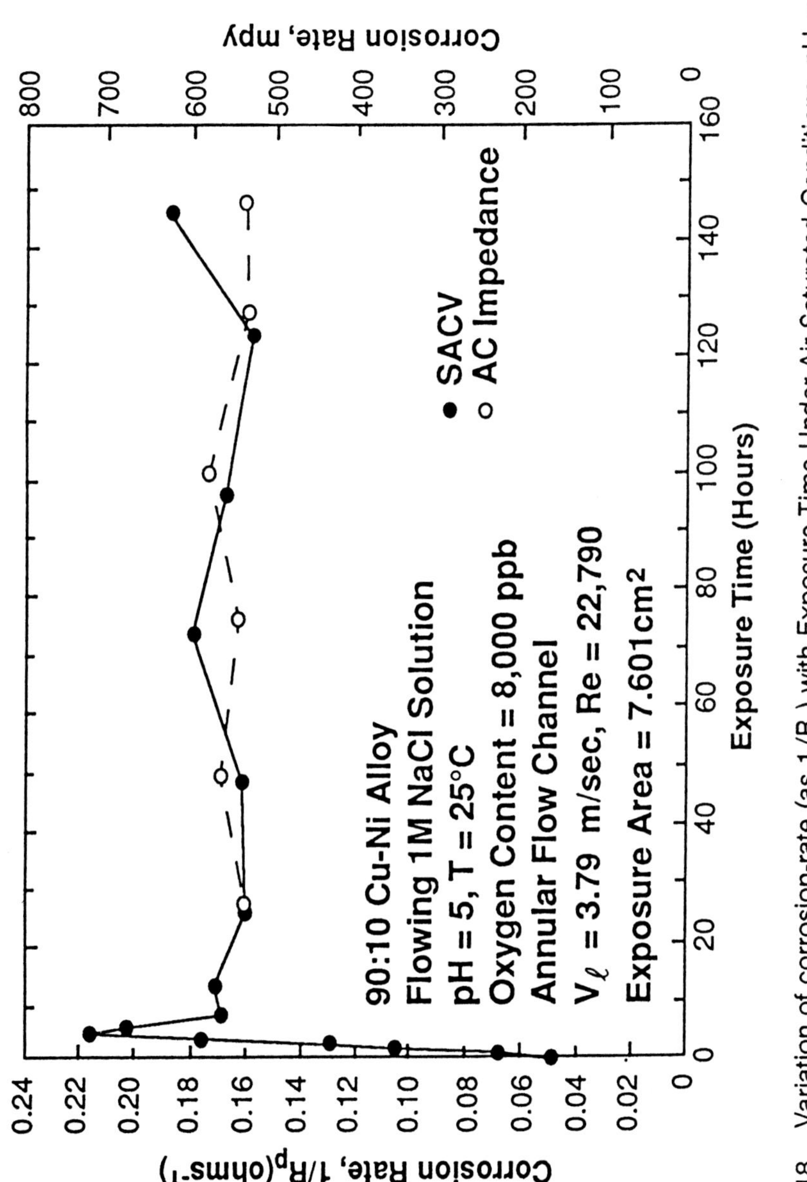

Figure 18   Variation of corrosion-rate (as $1/R_p$) with Exposure Time Under Air-Saturated Conditions, pH = 5, T = 25°C (77°F), for the Annular Flow Channel at Re = 22,790

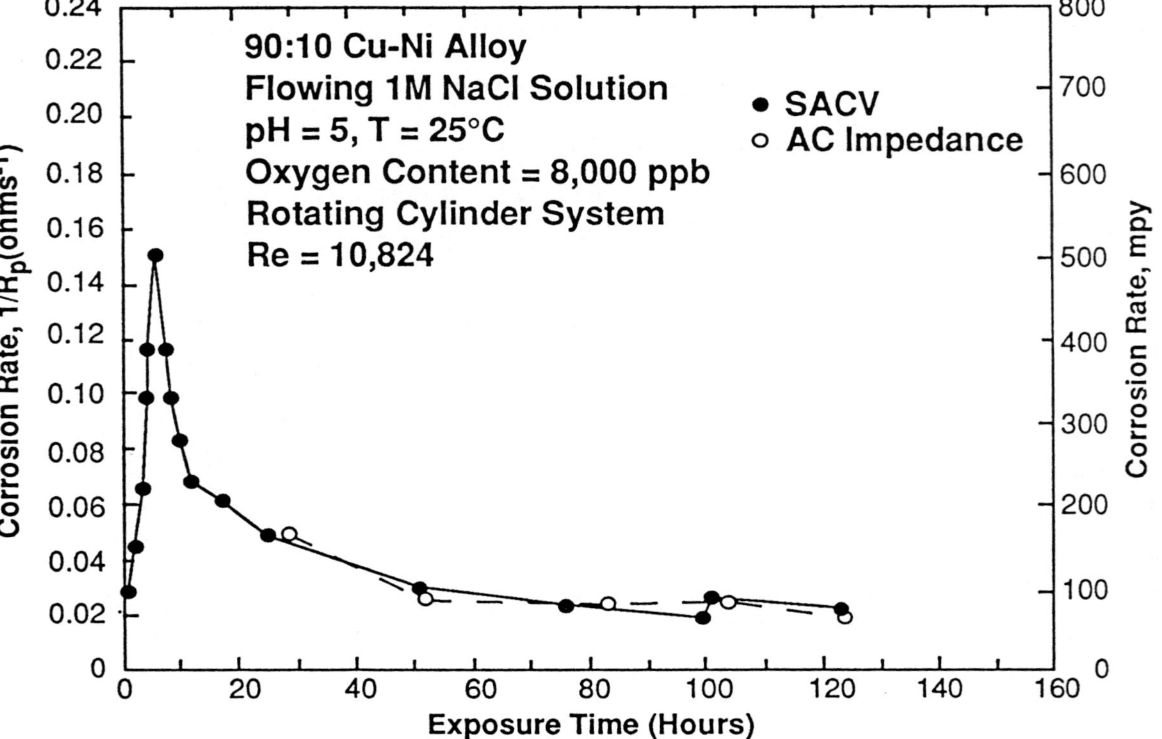

Figure 19   Variation of Corrosion Rate (as 1/R_p) with Exposure Time Under Air-Saturated Conditions, pH = 5, T = 25°C (77°F), for the Rotating Cylinder System at Re = 10,824

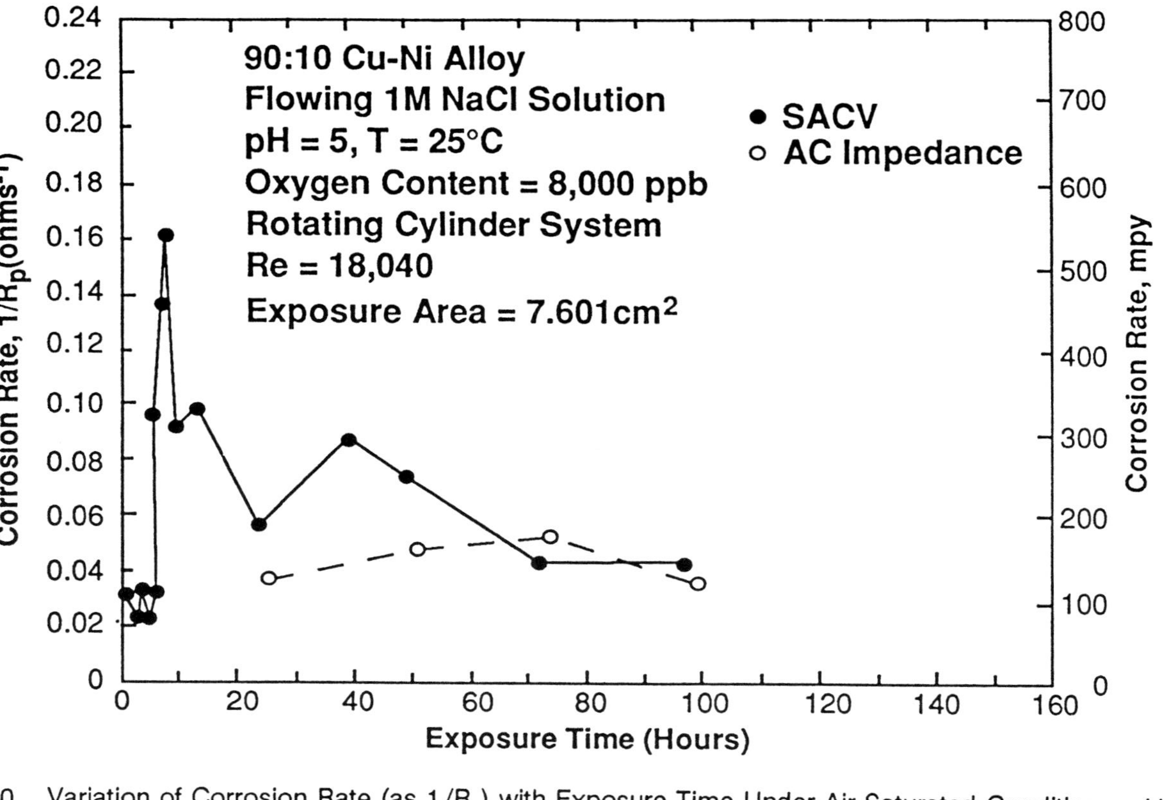

Figure 20   Variation of Corrosion Rate (as $1/R_p$) with Exposure Time Under Air-Saturated Conditions, pH = 5, T = 25°C (77°F), for the Rotating Cylinder System at Re = 18,040

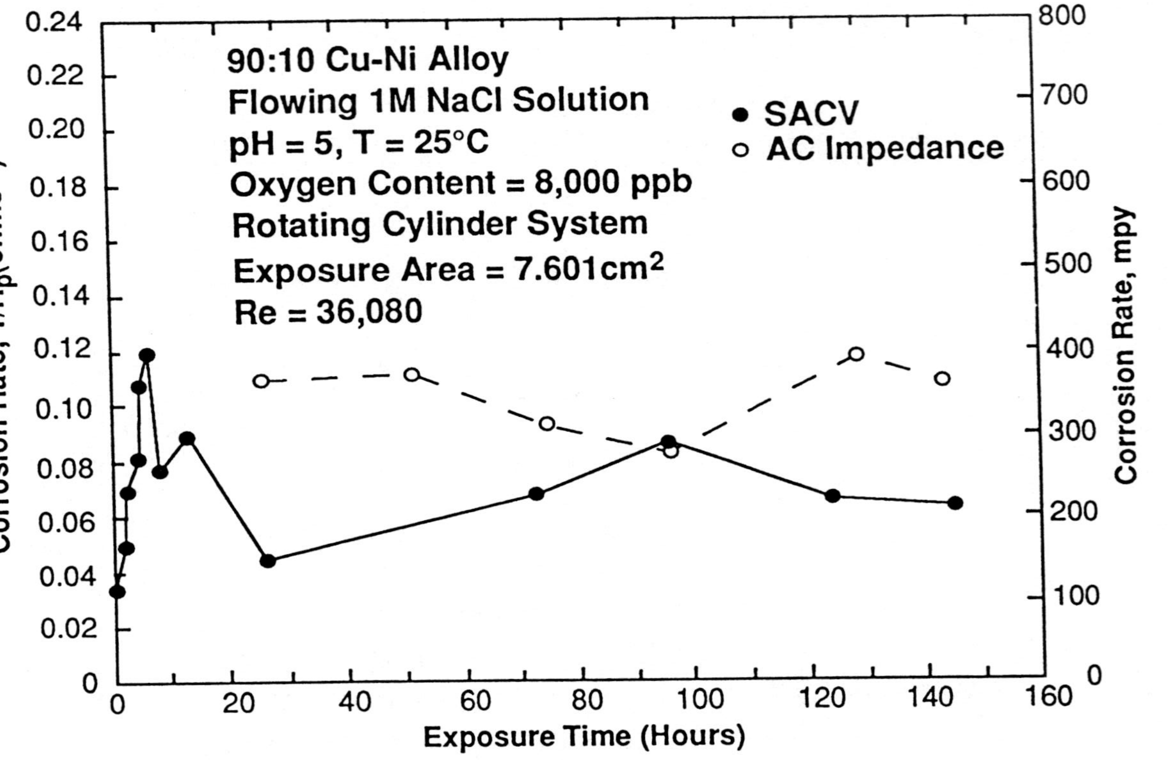

Figure 21   Variation of Corrosion Rate (as $1/R_p$) with Exposure Time Under Air-Saturated Conditions, pH = 5, T = 25°C (77°F), for the Rotating Cylinder System at Re = 36,080

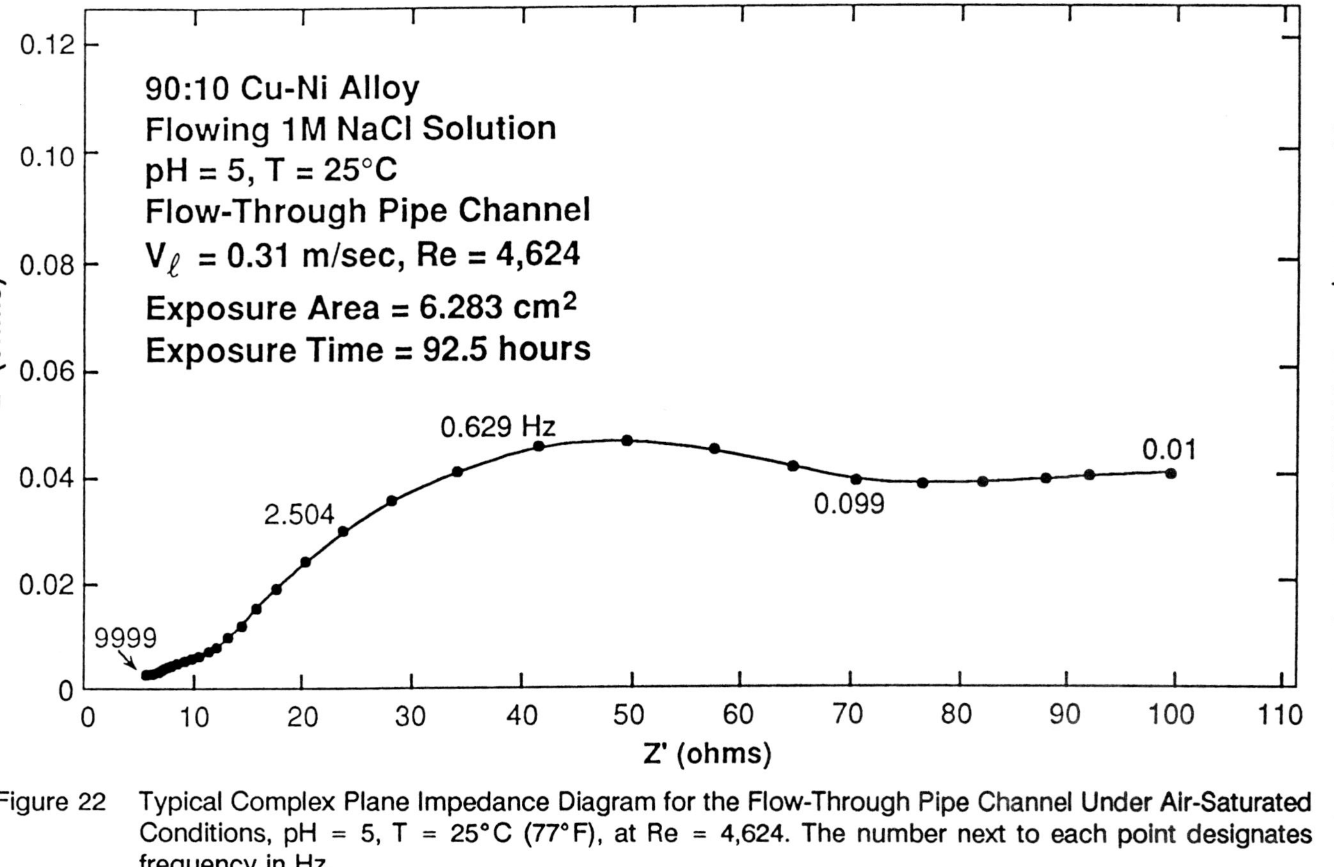

Figure 22   Typical Complex Plane Impedance Diagram for the Flow-Through Pipe Channel Under Air-Saturated Conditions, pH = 5, T = 25°C (77°F), at Re = 4,624. The number next to each point designates frequency in Hz.

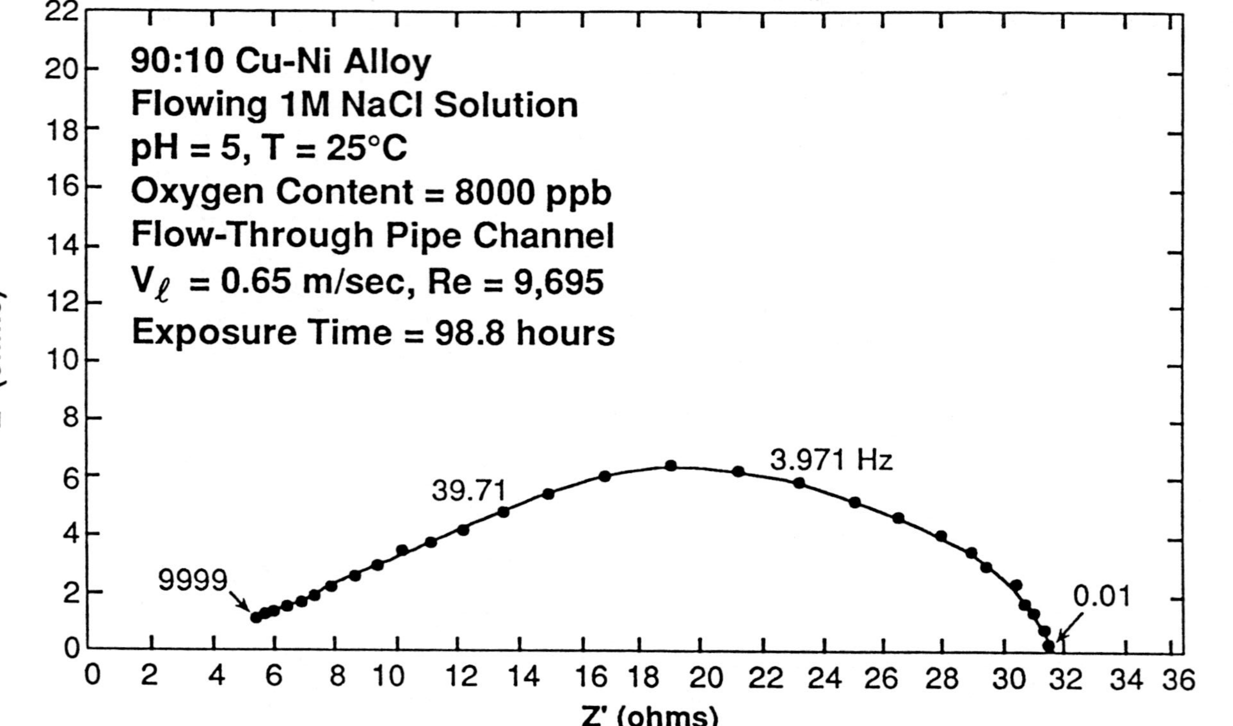

Figure 23   Typical Complex Plane Impedance Diagram for the Flow-Through Pipe Channel Under Air-Saturated Conditions, pH = 5, T = 25°C (77°F), at Re = 9,695. The number next to each point designates frequency in Hz.

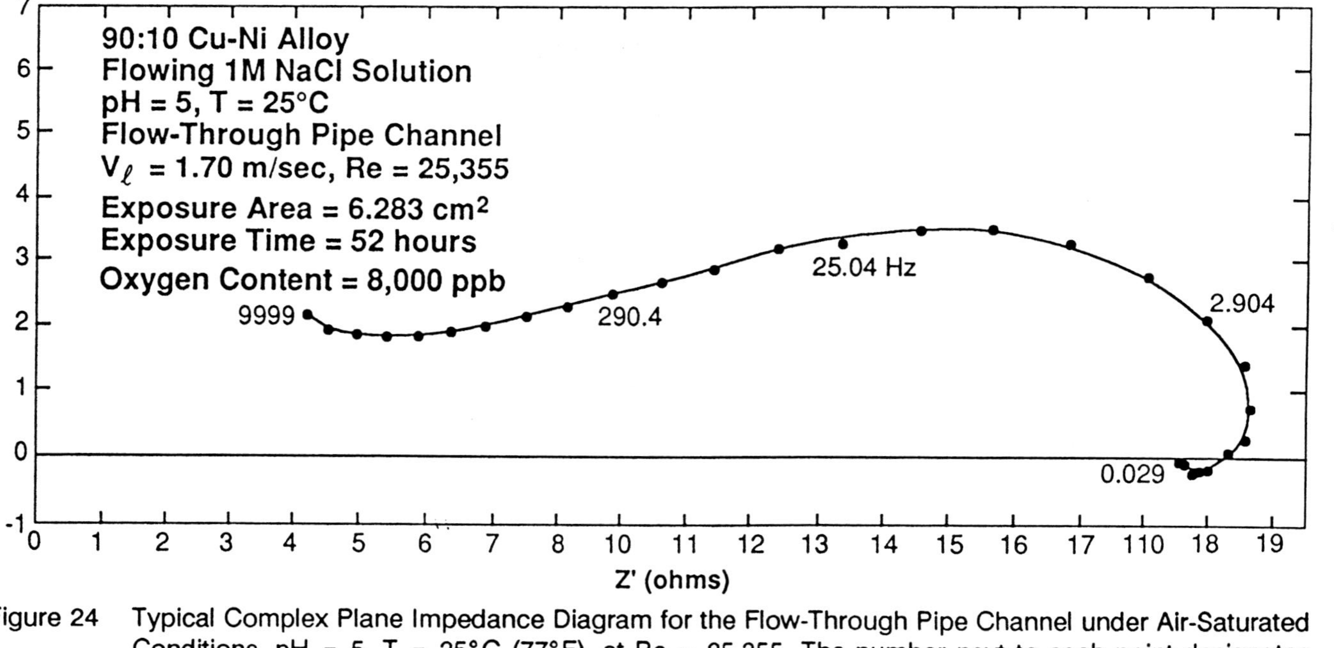

Figure 24   Typical Complex Plane Impedance Diagram for the Flow-Through Pipe Channel under Air-Saturated Conditions, pH = 5, T = 25°C (77°F), at Re = 25,355. The number next to each point designates frequency in Hz.

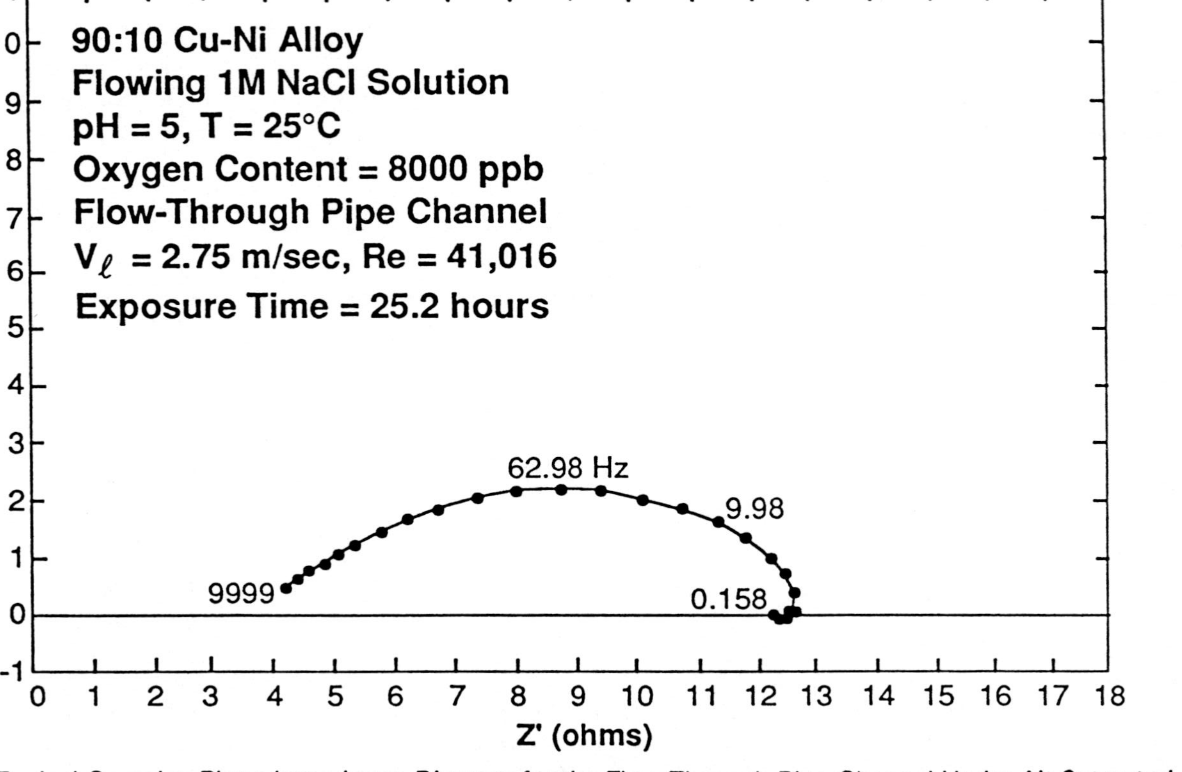

Figure 25   Typical Complex Plane Impedance Diagram for the Flow-Through Pipe Channel Under Air-Saturated Conditions, pH = 5, T = 25°C (77°F), at Re = 41,016. The number next to each point designates frequency in Hz.

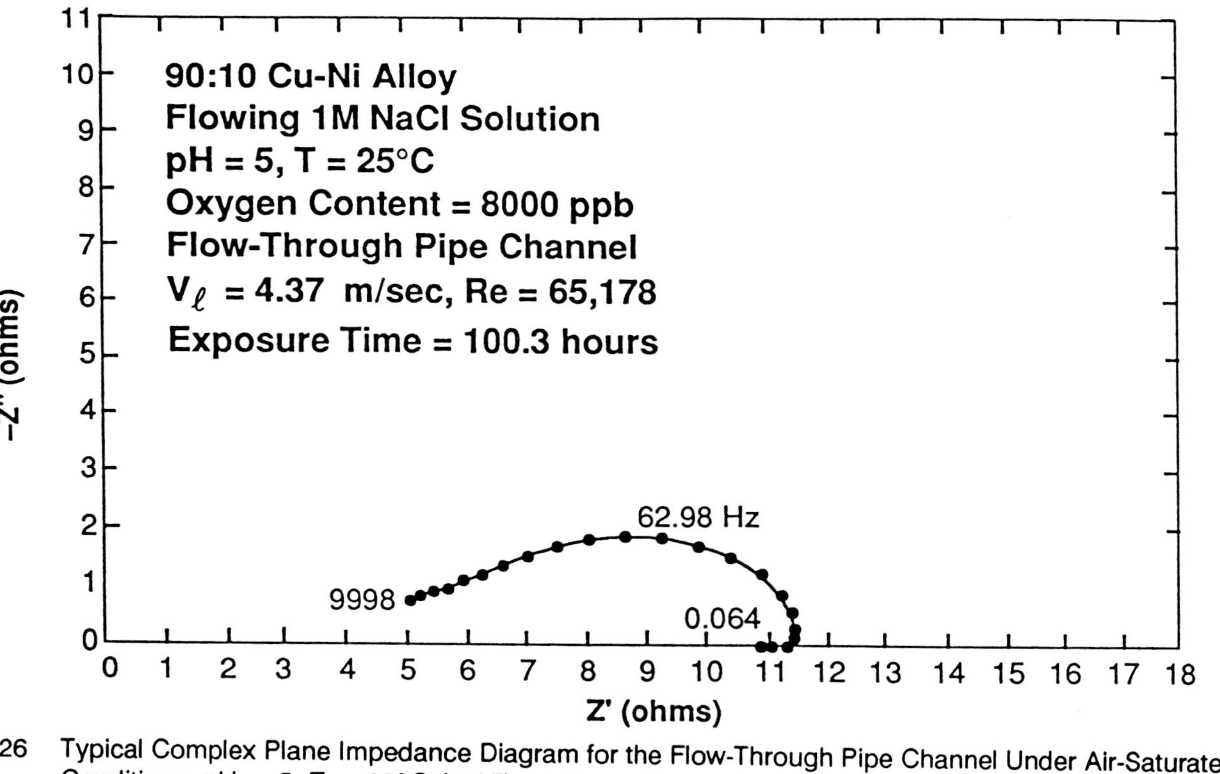

Figure 26   Typical Complex Plane Impedance Diagram for the Flow-Through Pipe Channel Under Air-Saturated Conditions, pH = 5, T = 25°C (77°F), at Re = 65,178. The number next to each point designates frequency in Hz.

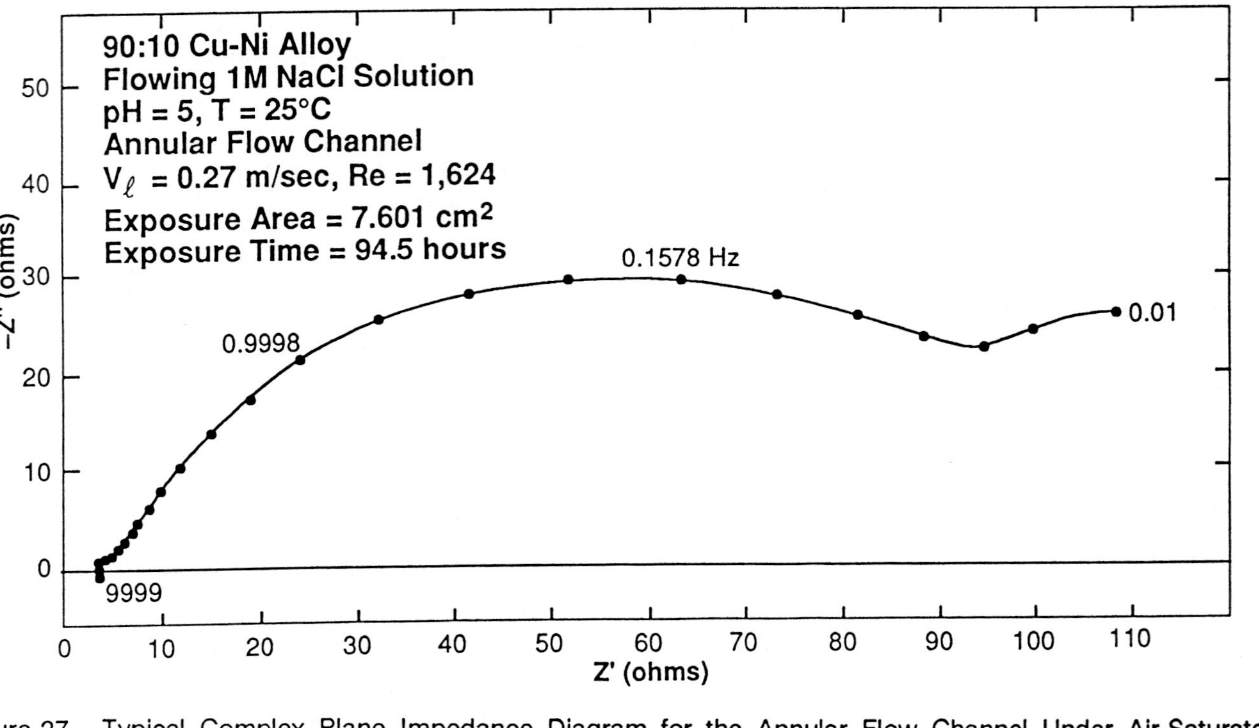

Figure 27   Typical Complex Plane Impedance Diagram for the Annular Flow Channel Under Air-Saturated Conditions, pH = 5, T = 25°C (77°F), at Re = 1,624. The number next to each point designates frequency in Hz.

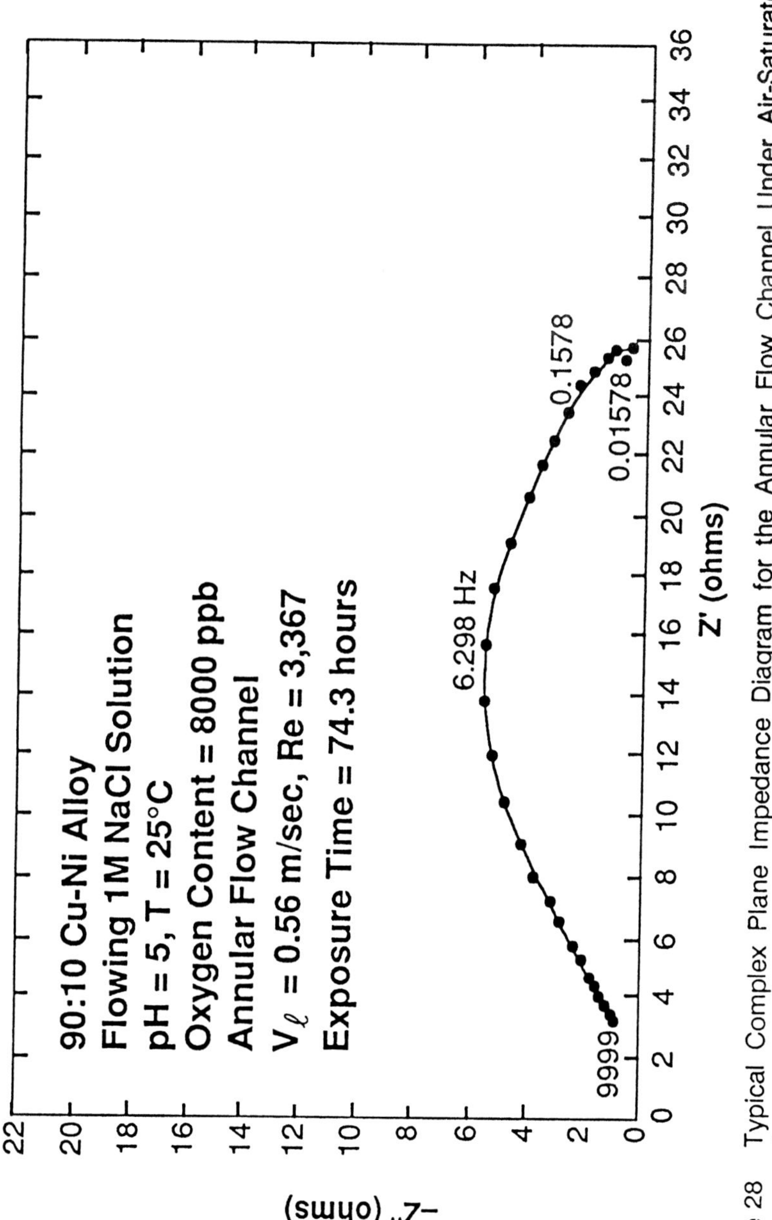

Figure 28   Typical Complex Plane Impedance Diagram for the Annular Flow Channel Under Air-Saturated Conditions, pH = 5, T = 25°C (77°F), at Re = 3,367. The number next to each point designates frequency in Hz.

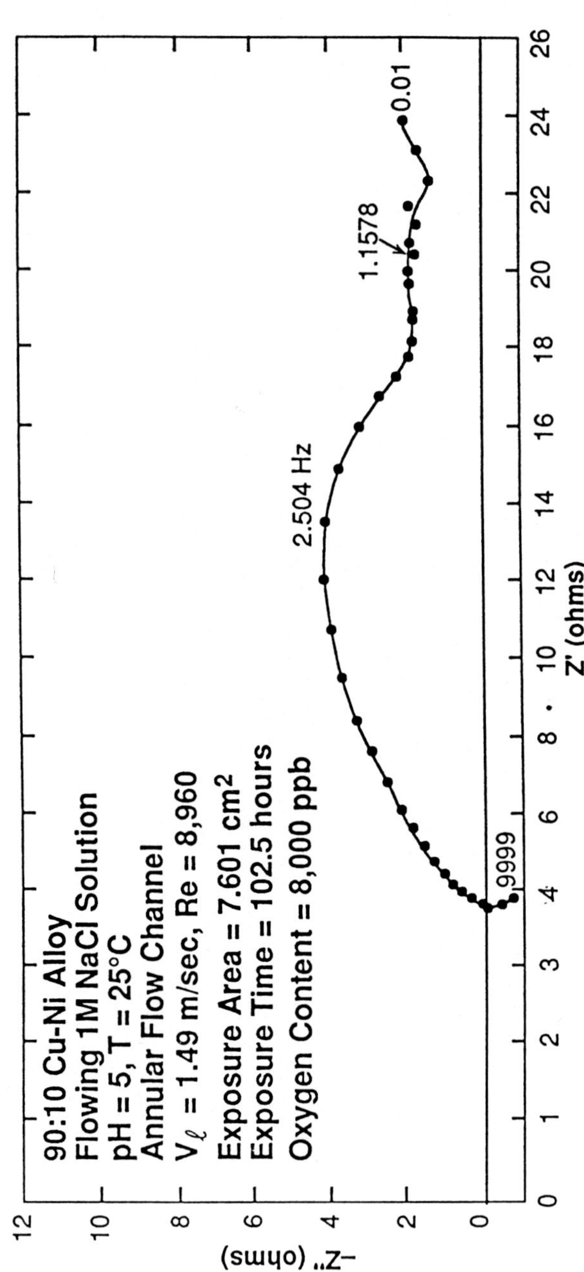

Figure 29   Typical Complex Plane Impedance Diagram for the Annular Flow Channel Under Air-Saturated Conditions, pH = 5, T = 25°C (77°F), at Re = 8,960. The number next to each point designates frequency in Hz.

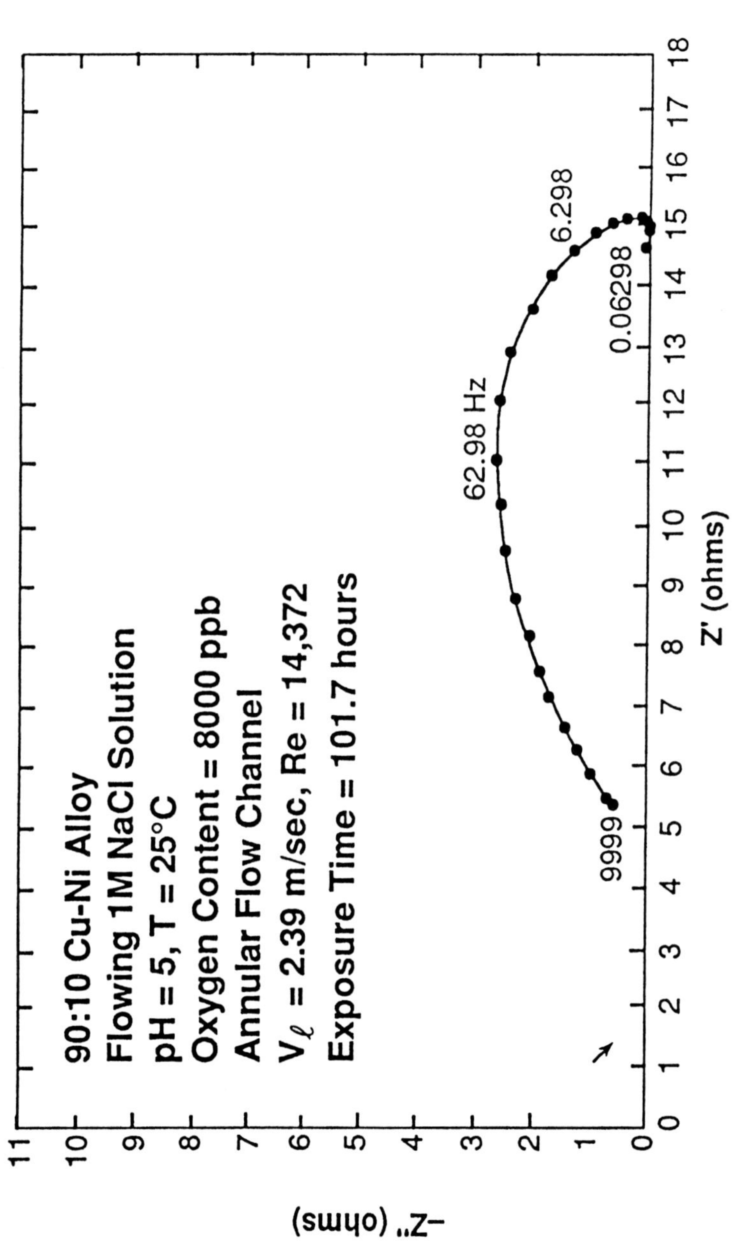

Figure 30   Typical Complex Plane Impedance Diagram for the Annular Flow Channel Under Air-Saturated Conditions, pH = 5, T = 25°C (77°F), at Re = 14,372. The number next to each point designates frequency in Hz.

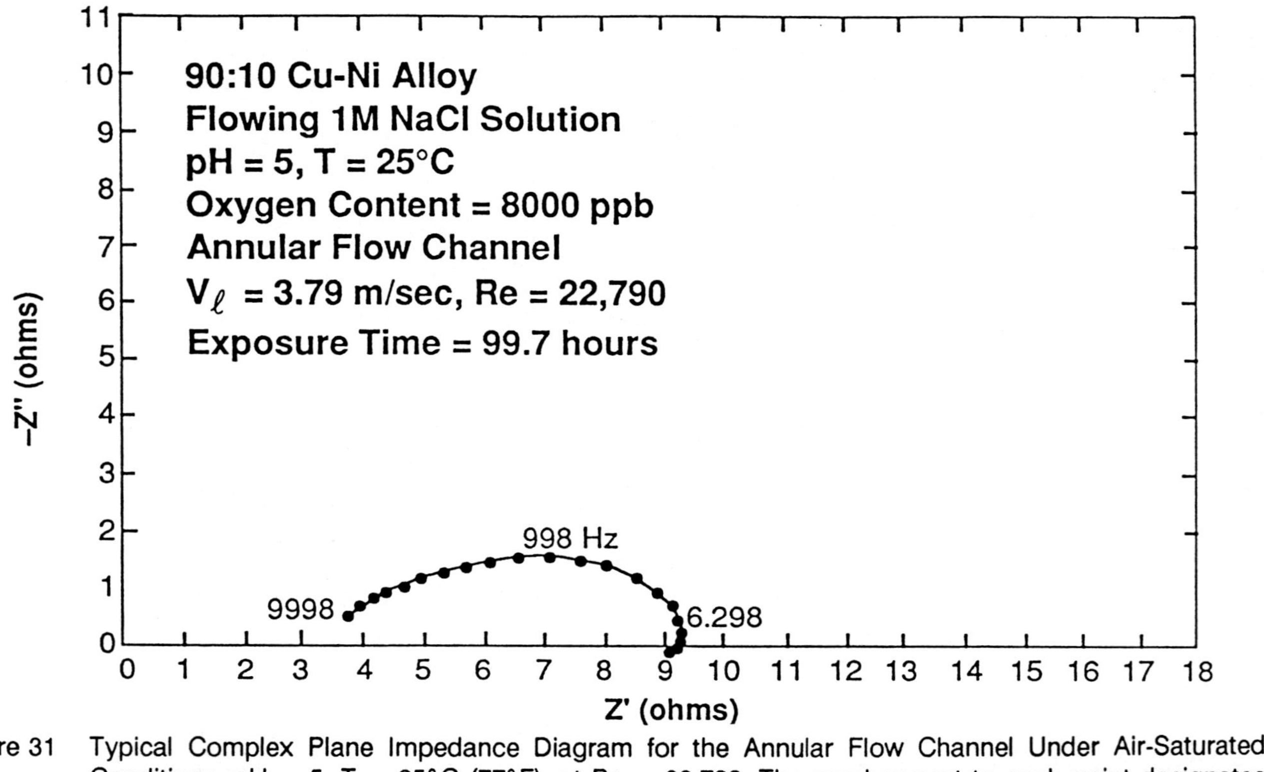

Figure 31   Typical Complex Plane Impedance Diagram for the Annular Flow Channel Under Air-Saturated Conditions, pH = 5, T = 25°C (77°F), at Re = 22,790. The number next to each point designates frequency in Hz.

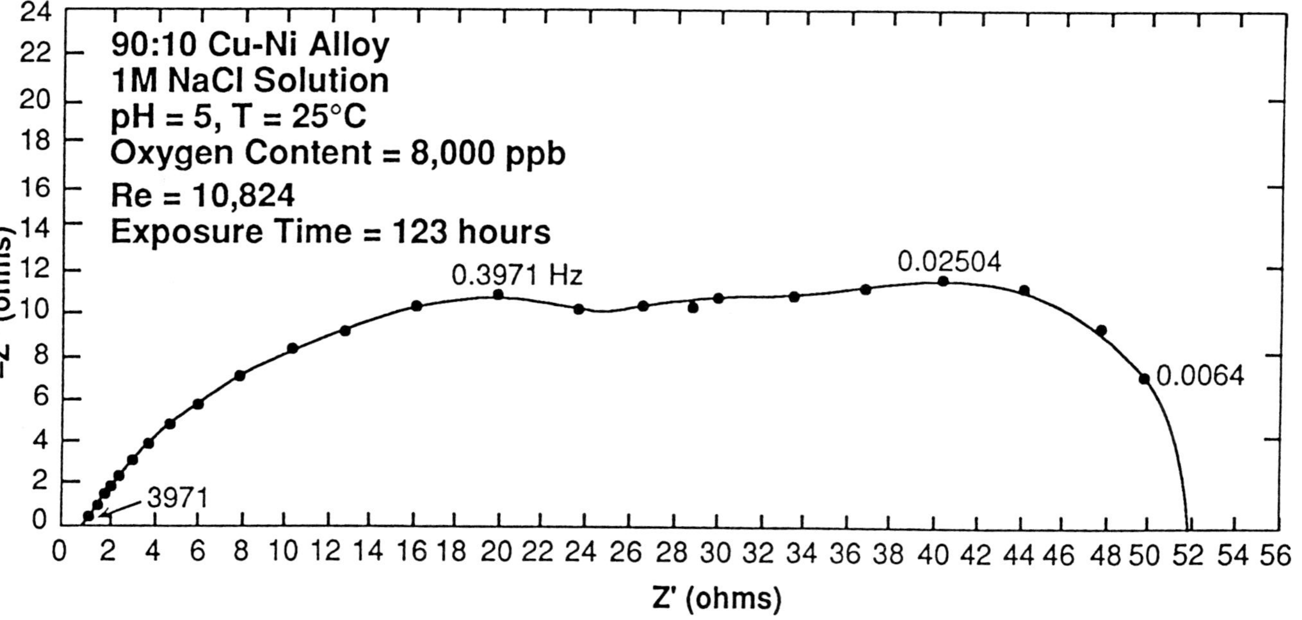

Figure 32   Typical Complex Plane Impedance Diagram for the Rotating Cylinder System Under Air-Saturated Conditions, pH = 5, T = 25°C (77°F), at Re = 10,824. The number next to each point designates frequency in Hz.

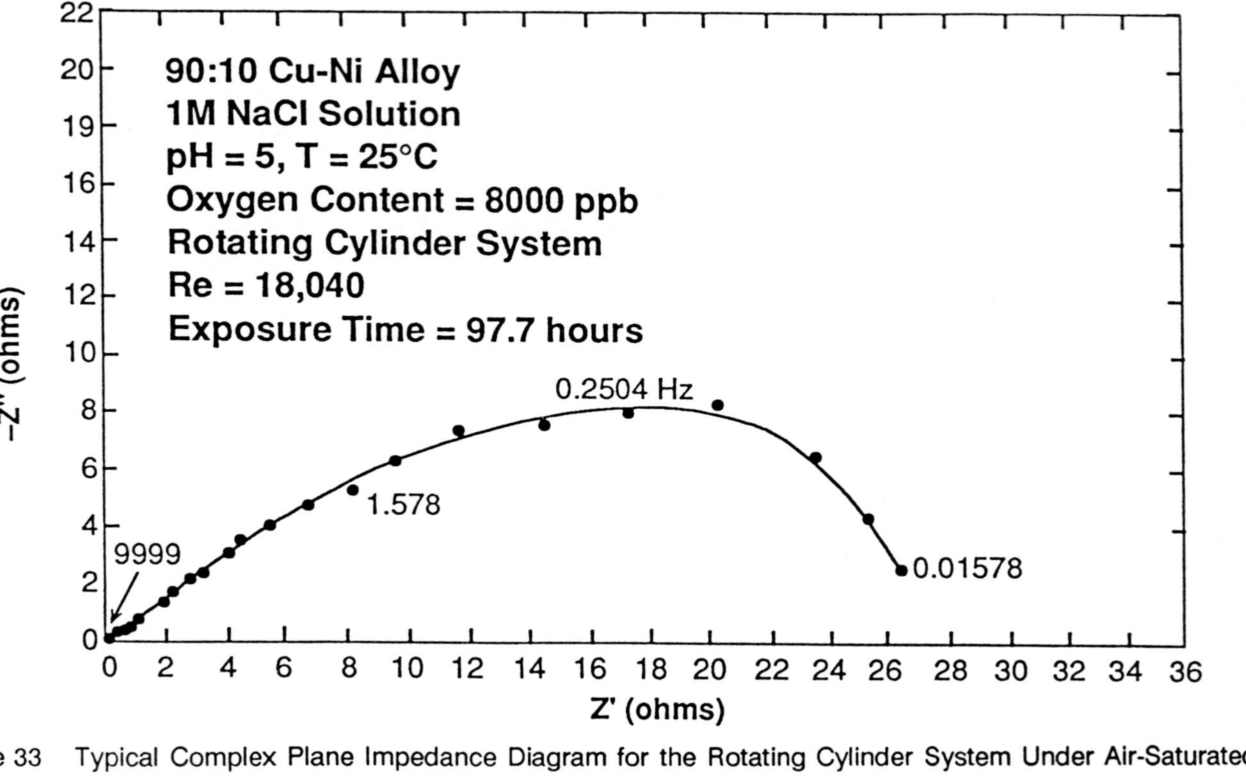

Figure 33   Typical Complex Plane Impedance Diagram for the Rotating Cylinder System Under Air-Saturated Conditions, pH = 5, T = 25°C (77°F), at Re = 18,040. The number next to each point designates frequency in Hz.

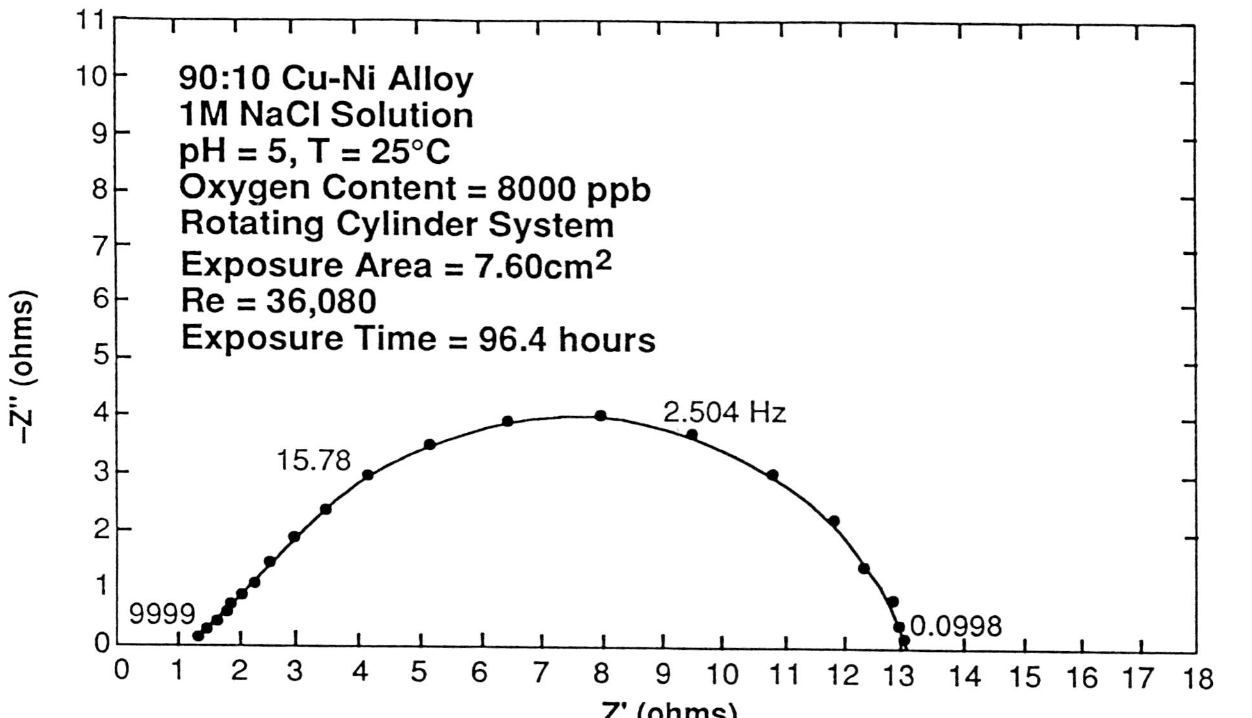

Figure 34   Typical Complex Plane Impedance Diagram for the Rotating Cylinder System Under Air-Saturated Conditions, pH = 5, T = 25°C (77°F), at Re = 36,080. The number next to each point designates frequency in Hz.

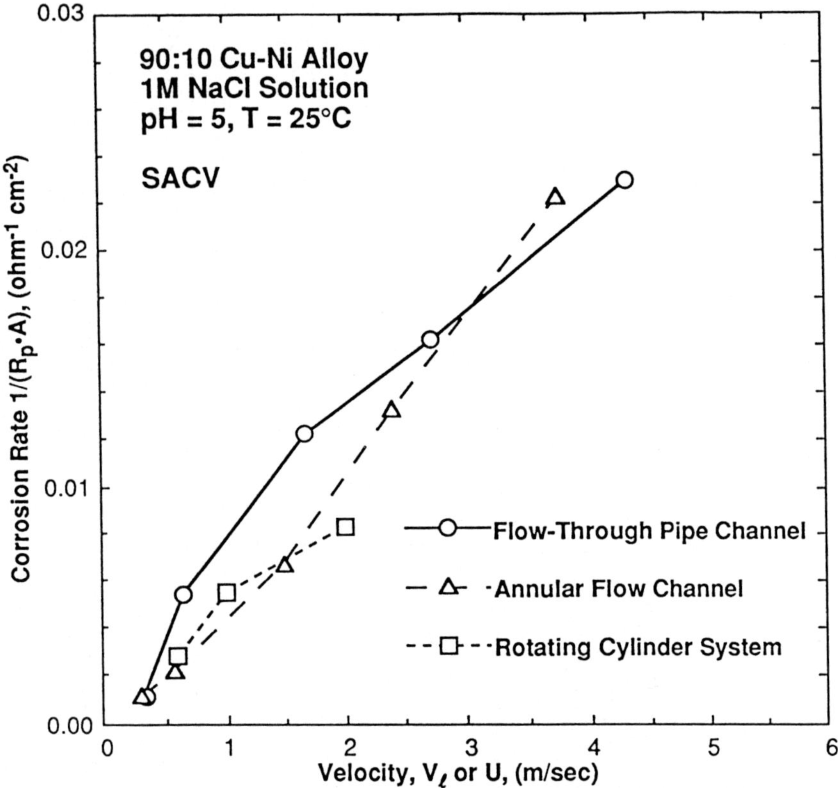

Figure 35   Variation of Corrosion Rate (as $1/R_pA$) with Velocity Under Air-Saturated Conditions, pH = 5, T = 25°C (77°F)

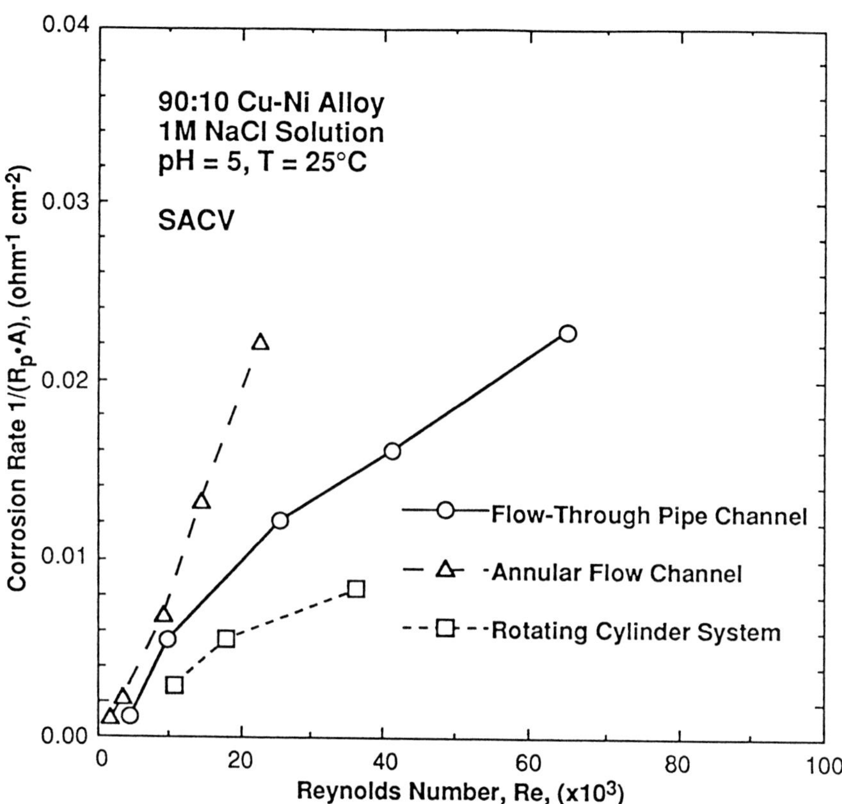

Figure 36   Variation of Corrosion Rate (as $1/R_pA$) with Reynolds Number Under Air-Saturated Conditions, pH = 5, T = 25°C (77°F)

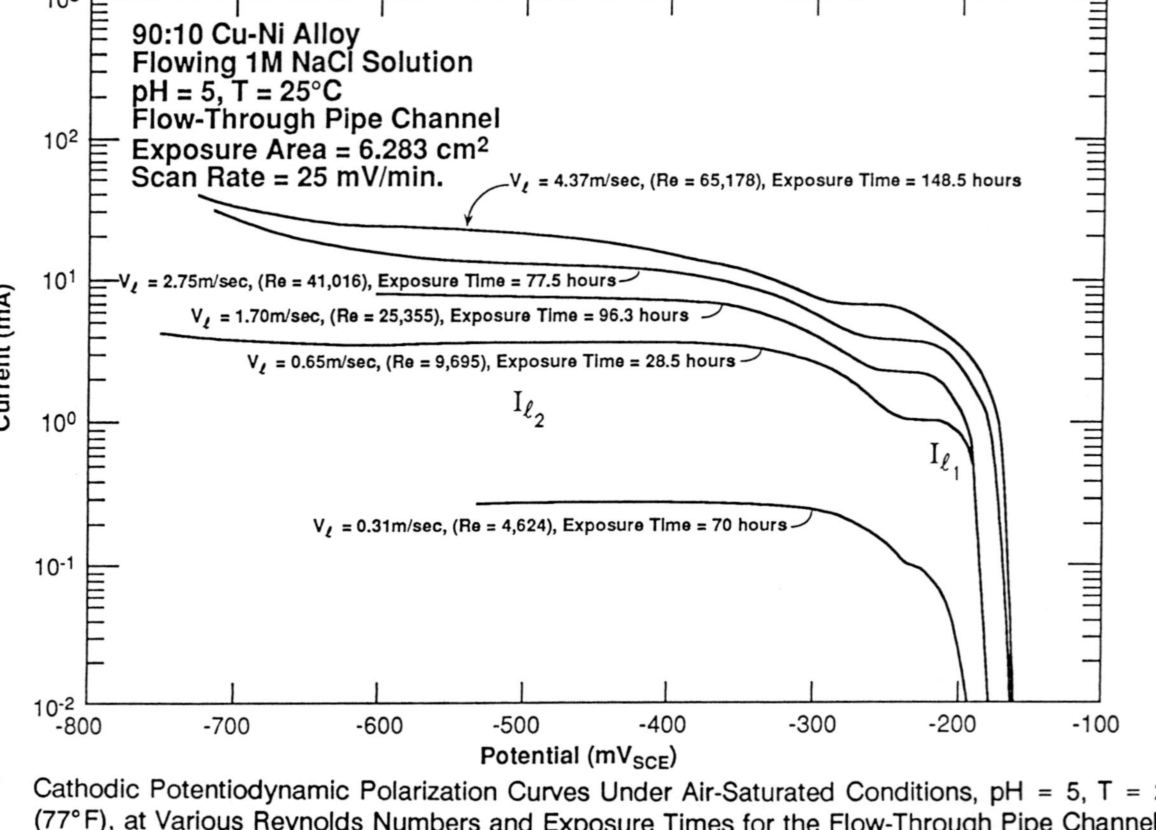

Figure 37   Cathodic Potentiodynamic Polarization Curves Under Air-Saturated Conditions, pH = 5, T = 25°C (77°F), at Various Reynolds Numbers and Exposure Times for the Flow-Through Pipe Channel

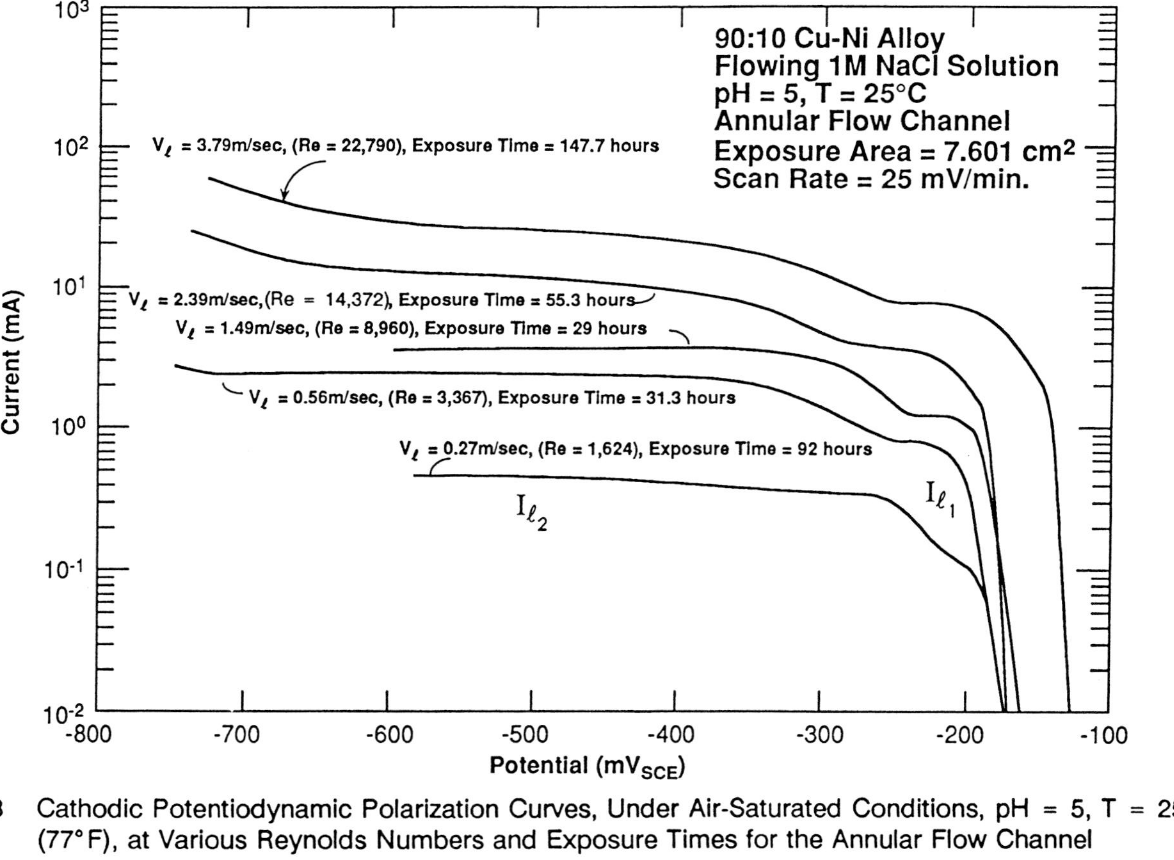

Figure 38   Cathodic Potentiodynamic Polarization Curves, Under Air-Saturated Conditions, pH = 5, T = 25°C (77°F), at Various Reynolds Numbers and Exposure Times for the Annular Flow Channel

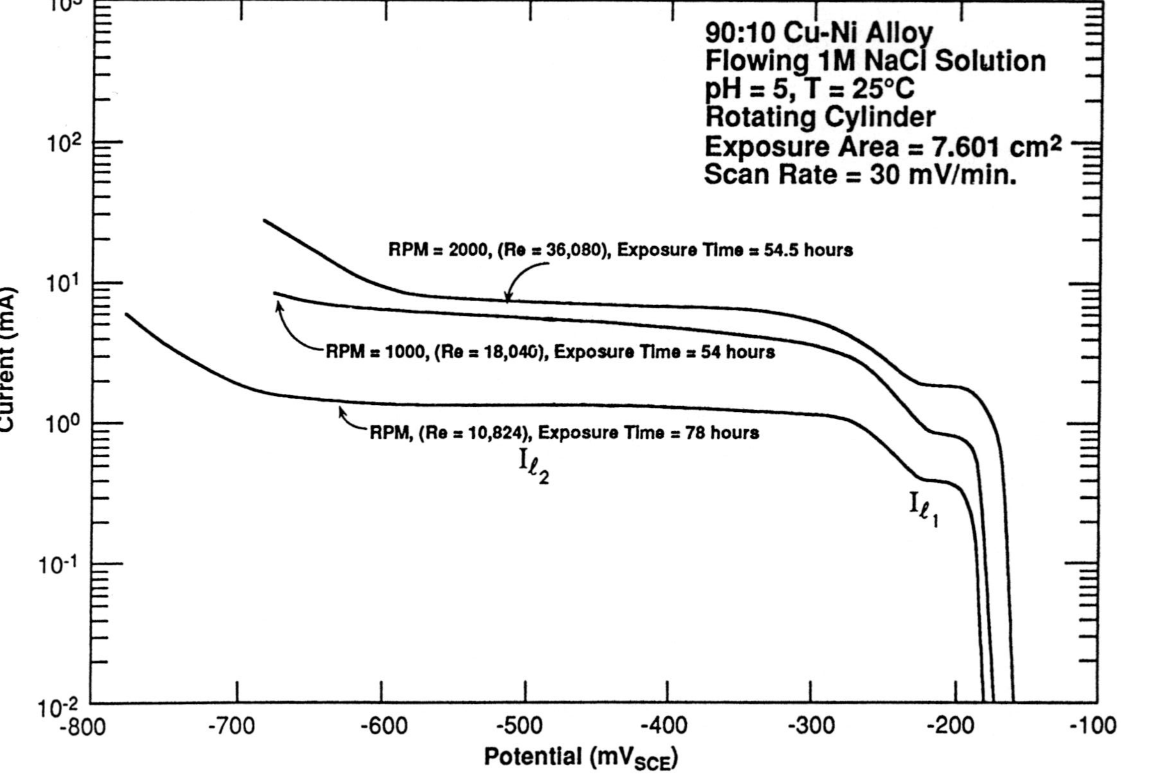

Figure 39   Cathodic Potentiodynamic Polarization Curves Under Air-Saturated Conditions, pH = 5, T = 25°C (77°F), at Various Reynolds Numbers and Exposure Times for the Rotating Cylinder System

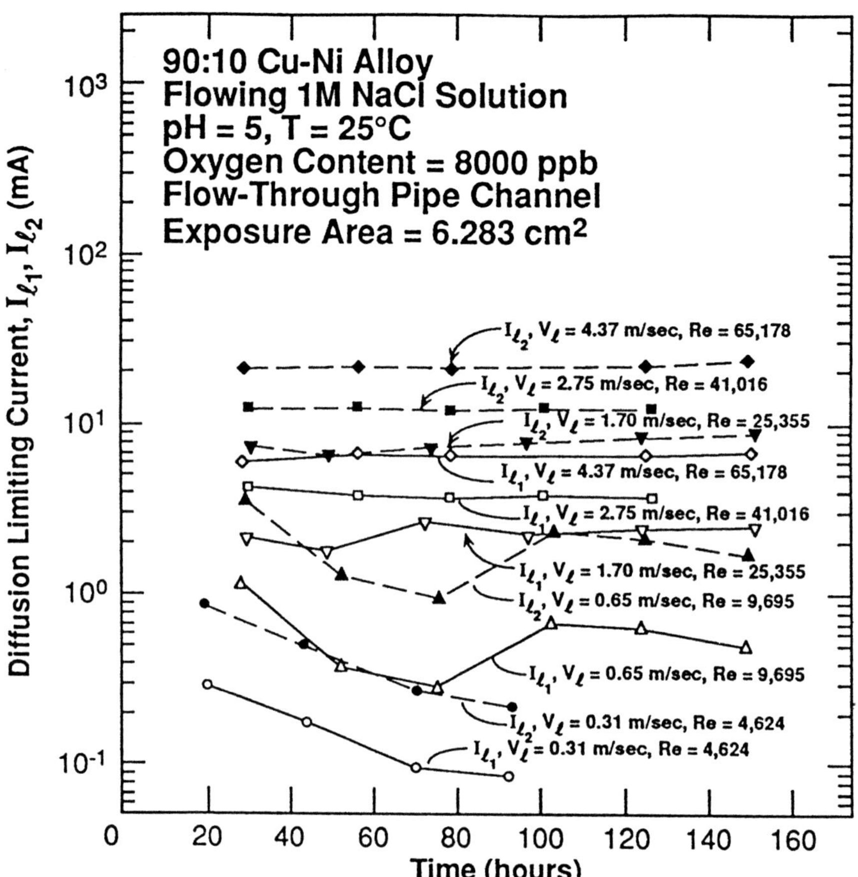

Figure 40   Variation of Diffusion Limiting Currents, I_{ℓ_1} and I_{ℓ_2}, with Exposure Time Under Air-Saturated Conditions, pH = 5, T = 25°C (77°F), at Different Reynolds Numbers for the Flow-Through Pipe Channel

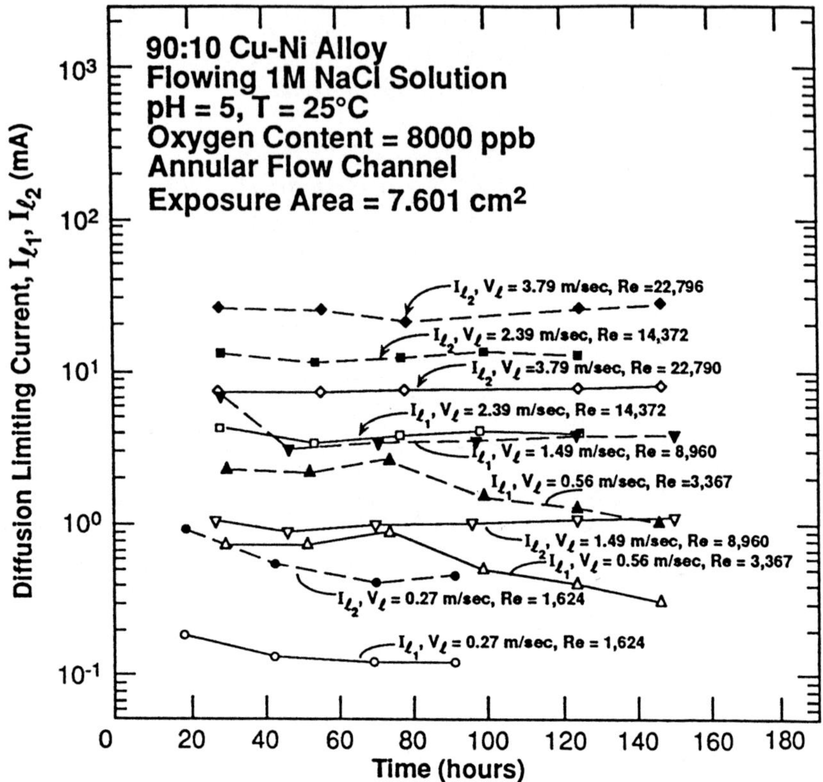

Figure 41   Variation of Diffusion Limiting Currents, I_{ℓ_1} and I_{ℓ_2}, with Exposure Time Under Air-Saturated Conditions, pH = 5, T = 25°C (77°F), at Different Reynolds Numbers for the Annular Flow Channel

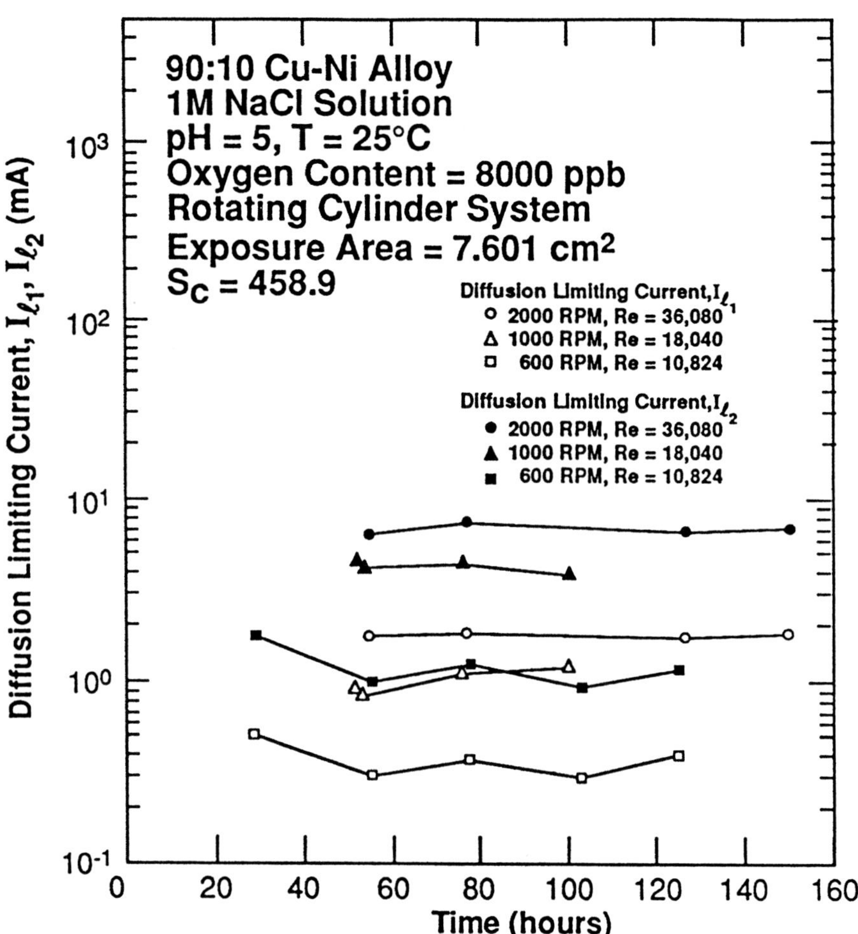

Figure 42   Variation of Diffusion Limiting Currents, I_{ℓ_1} and I_{ℓ_2}, with Exposure Time Under Air-Saturated Conditions, pH = 5, T = 25°C (77°F), at Different Reynolds Numbers for the Rotating Cylinder System

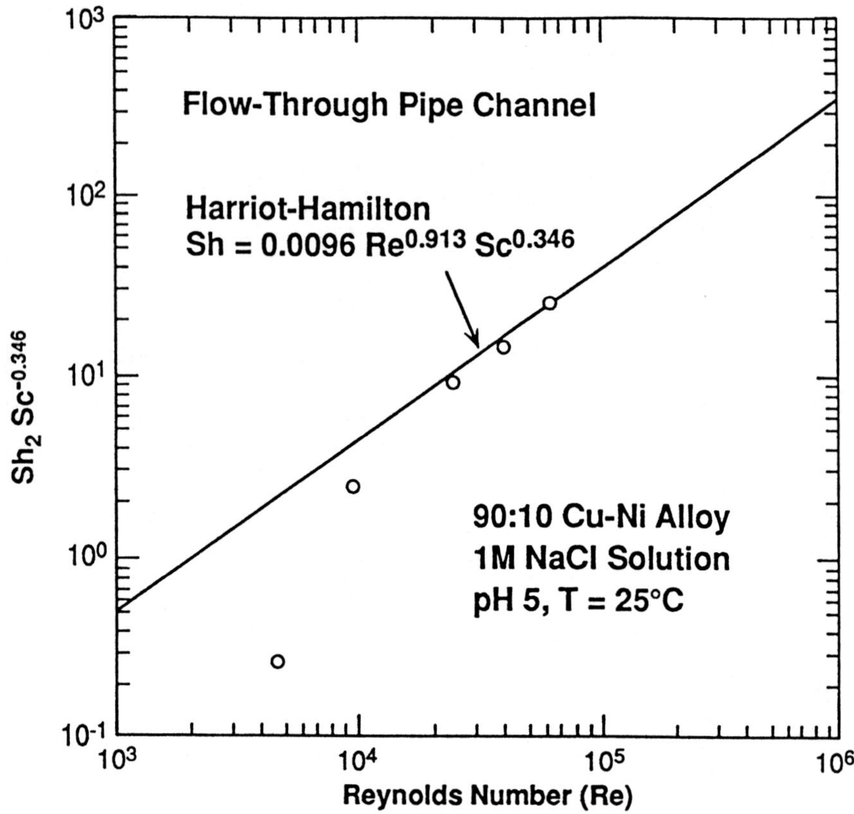

Figure 43   Comparison Between Experimentally Derived Mass Transfer Correlations for the Flow-Through Pipe Channel and Previously Published Data

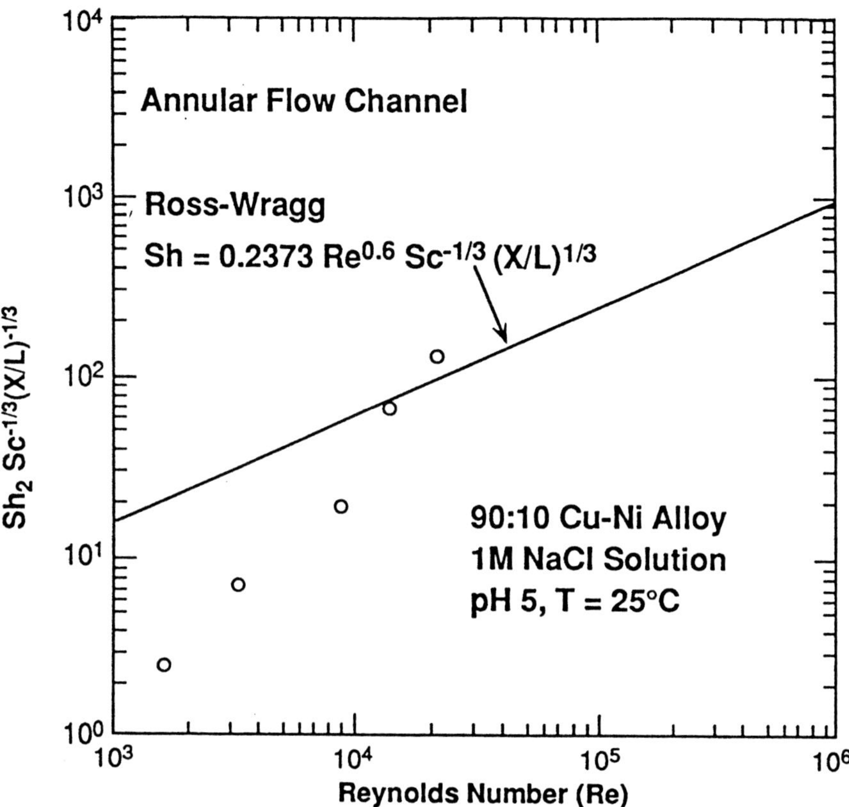

Figure 44   Comparison Between Experimentally Derived Mass Transfer Correlations for the Annular Flow Channel and Previously Published Data

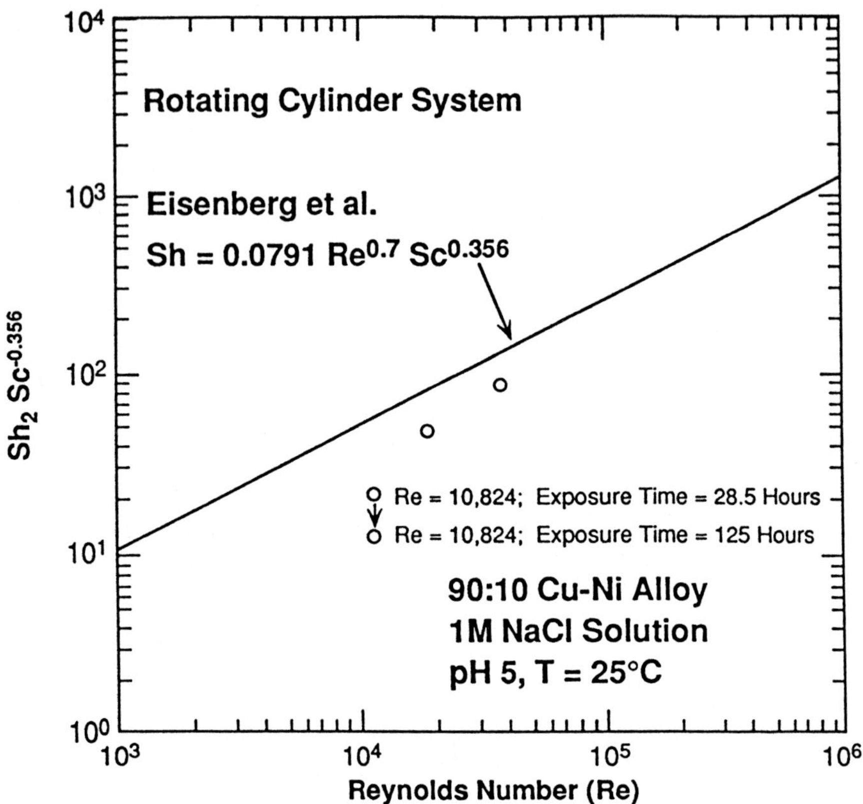

Figure 45   Comparison Between Experimentally Derived Mass Transfer Correlations for the Rotating Cylinder System and Previously Published Data

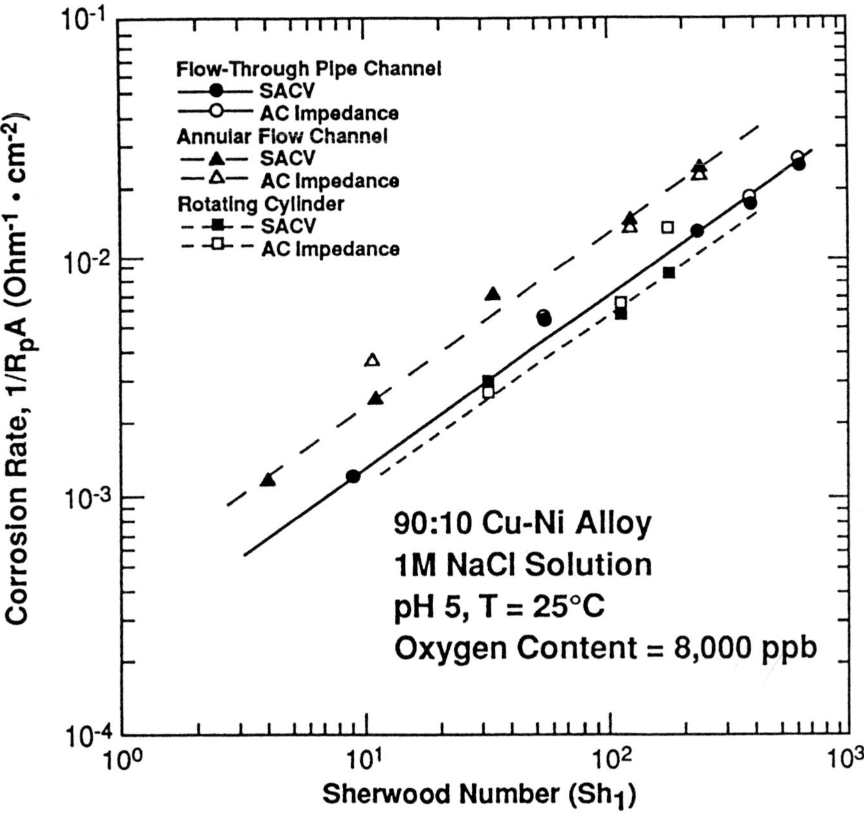

Figure 46   Variation of Corrosion Rate (as $1/R_pA$) Obtained Under Air-Saturated Conditions, pH = 5, T = 25°C (77°F), as a Function of Sh$_1$

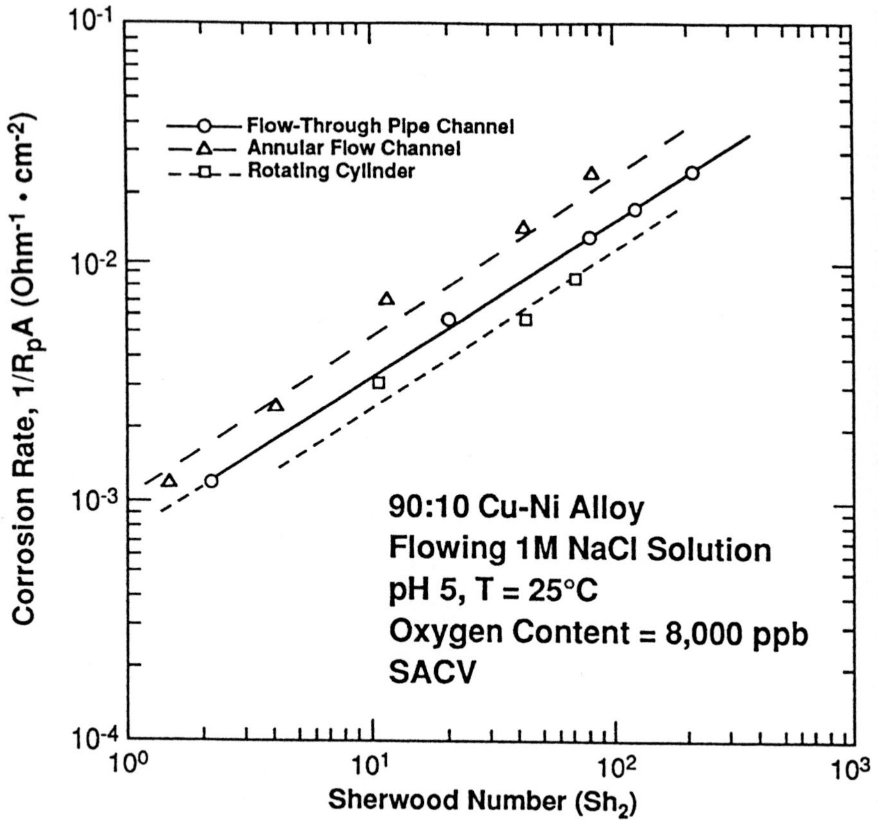

Figure 47   Variation of Corrosion Rate (as 1/R$_p$A) Obtained Under Air-Saturated Conditions, pH = 5, T = 25°C (77°F), as a Function of Sh$_2$

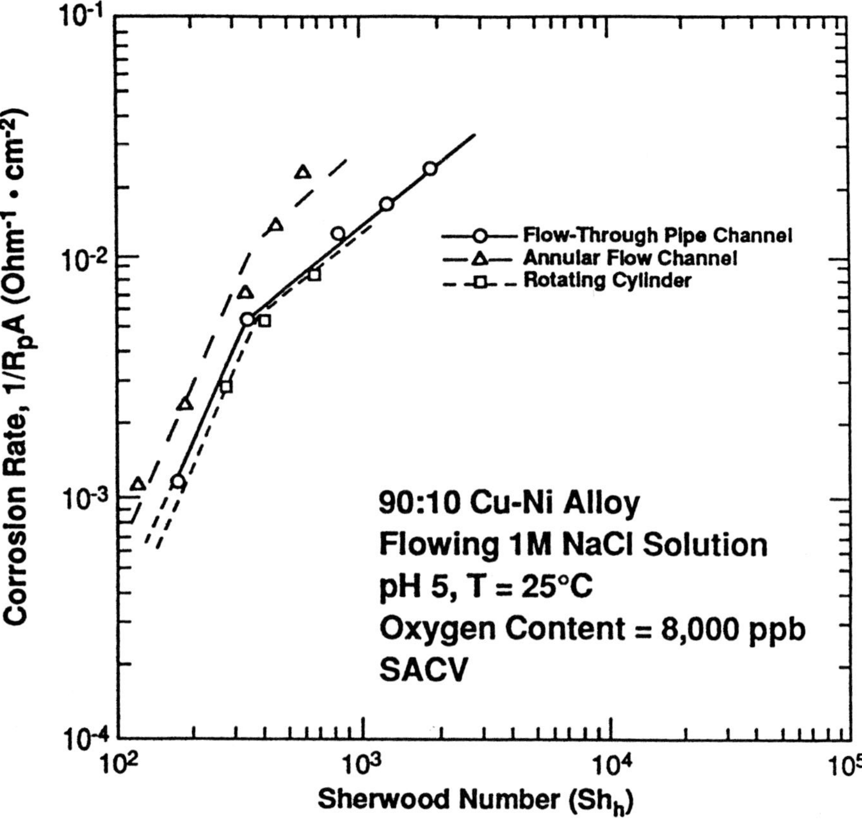

Figure 48 Variation of Corrosion Rate (as $1/R_pA$) Obtained Under Air-Saturated Conditions, pH = 5, T = 25°C (77°F), as a Function of Sh_h

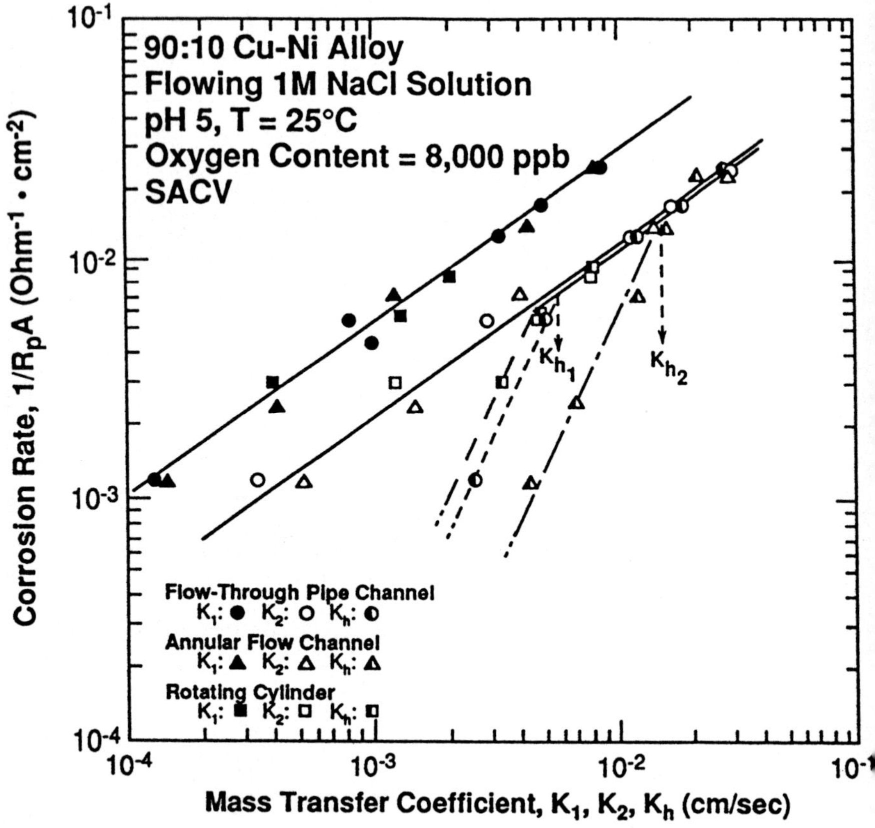

Figure 49   Variation of Corrosion Rate (as $1/R_pA$) Obtained Under Air-Saturated Conditions, pH = 5, T = 25°C (77°F), as a Function of Mass Transfer Coefficients, K_1, K_2, and K_h

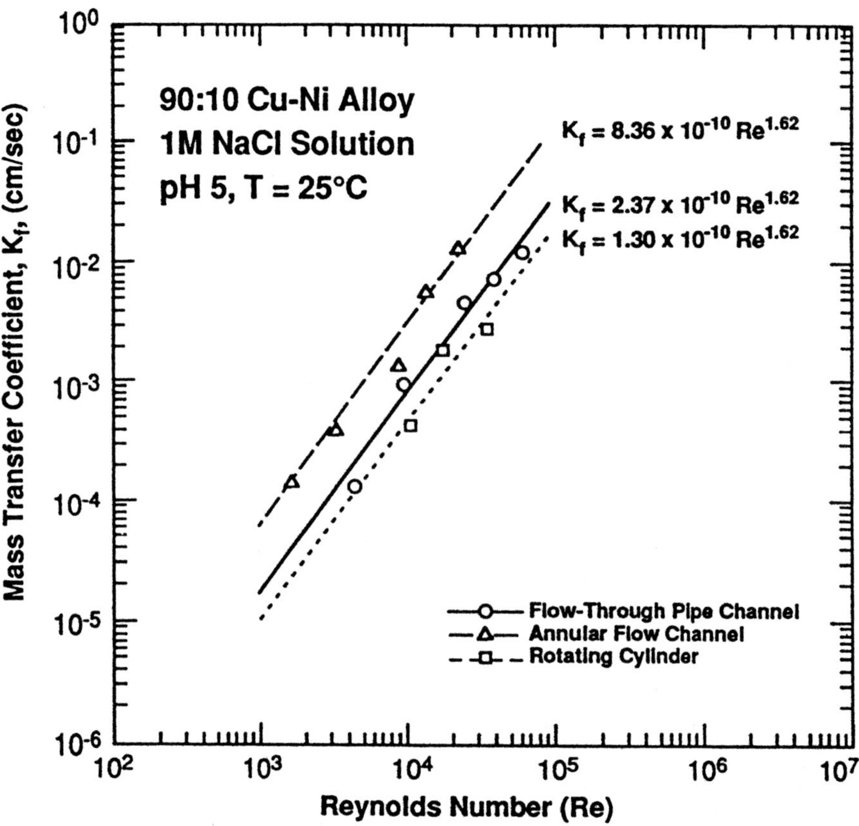

Figure 50   Variation of Mass Transfer Coefficient in the Surface Film, K_f, with Reynolds Number Under Air-Saturated Conditions, pH = 5, T = 25°C (77°F)

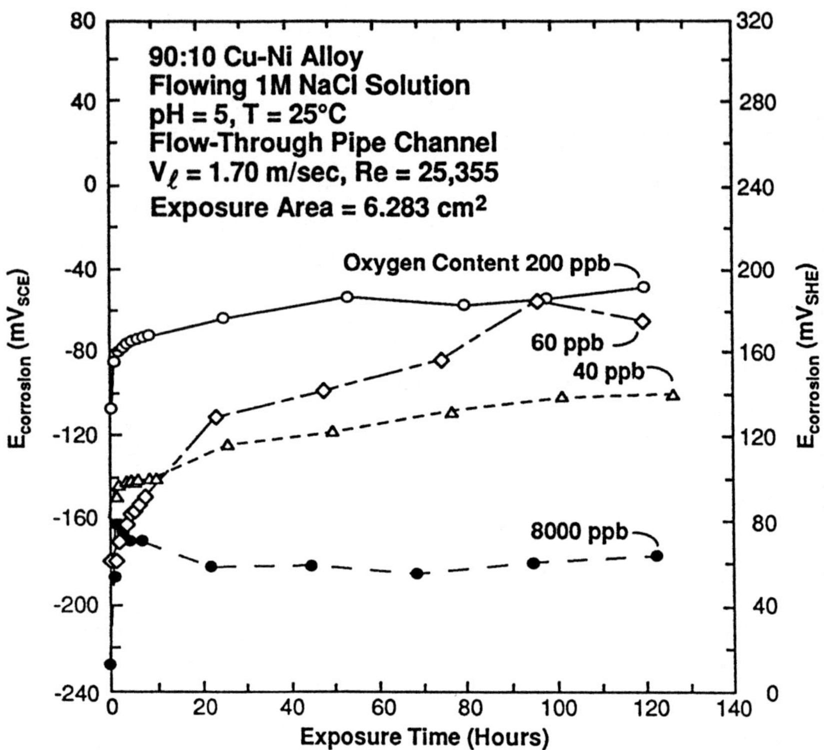

Figure 51   Variation of Corrosion Potential with Exposure Time for Different Dissolved Oxygen Concentrations at Re = 25,355, pH = 5, T = 25°C (77°F), for the Flow-Through Pipe Channel

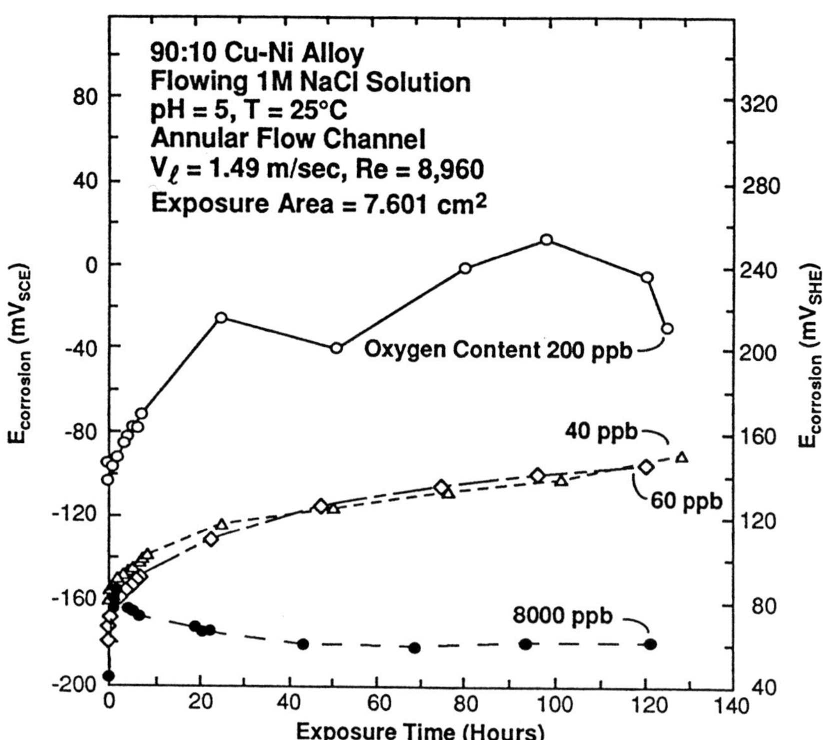

Figure 52   Variation of Corrosion Potential with Exposure Time for Different Dissolved Oxygen Concentrations at Re = 8,960, pH = 5, T = 25°C (77°F), for the Annular Flow Channel

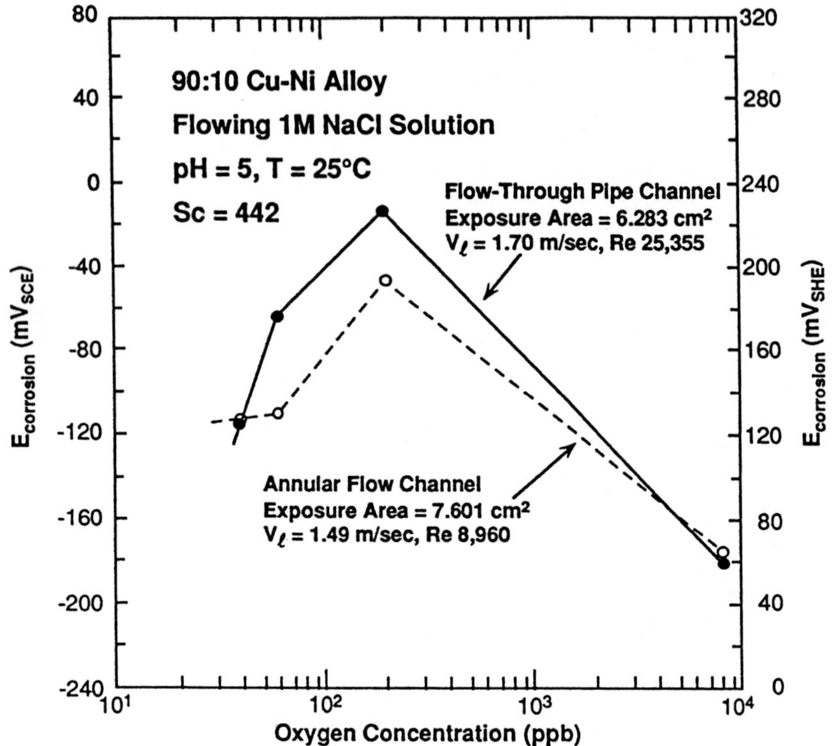

Figure 53   Curves of Corrosion Potential vs Dissolved Oxygen Concentration

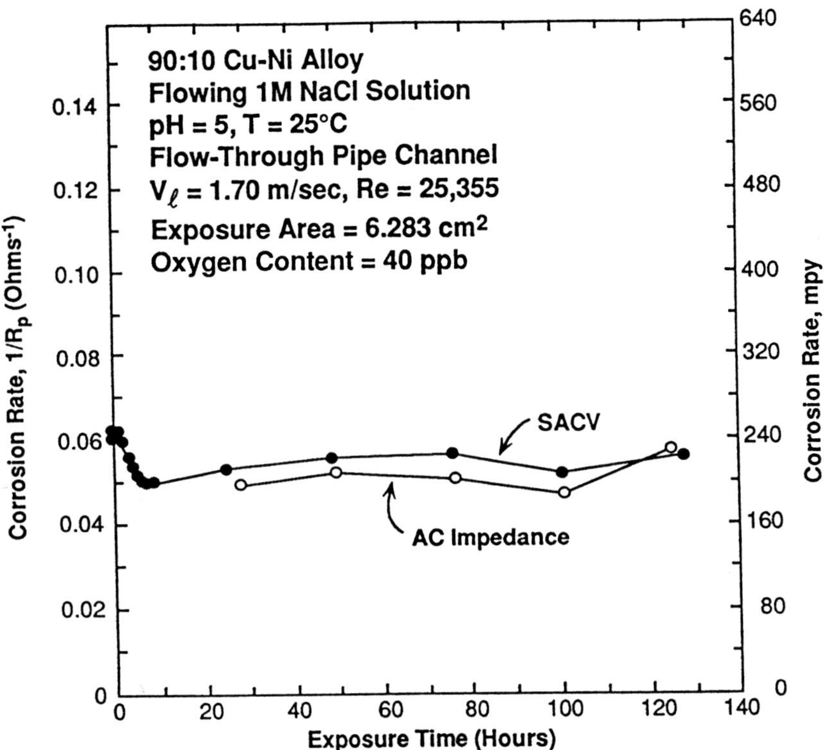

Figure 54   Variation of Corrosion Rate (as $1/R_p$) with Exposure Time for the Flow-Through Pipe Channel at Re = 25,355, pH = 5, T = 25°C (77°F), [O_2] = 40 ppb

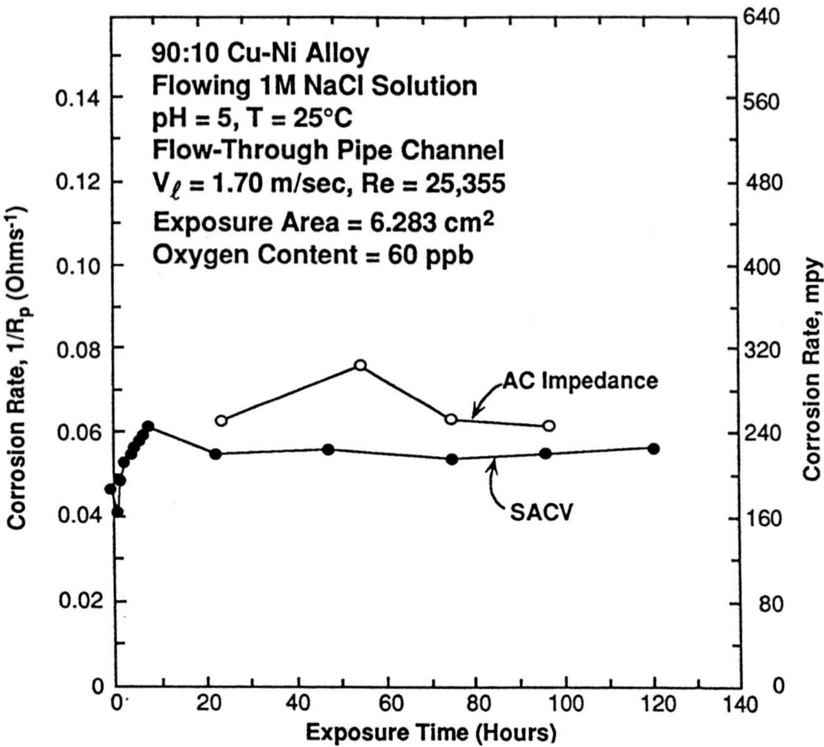

Figure 55   Variation of Corrosion Rate (as $1/R_p$) with Exposure Time for the Flow-Through Pipe Channel at Re = 25,355, pH = 5, T = 25°C (77°F), $[0_2]$ = 60 ppb

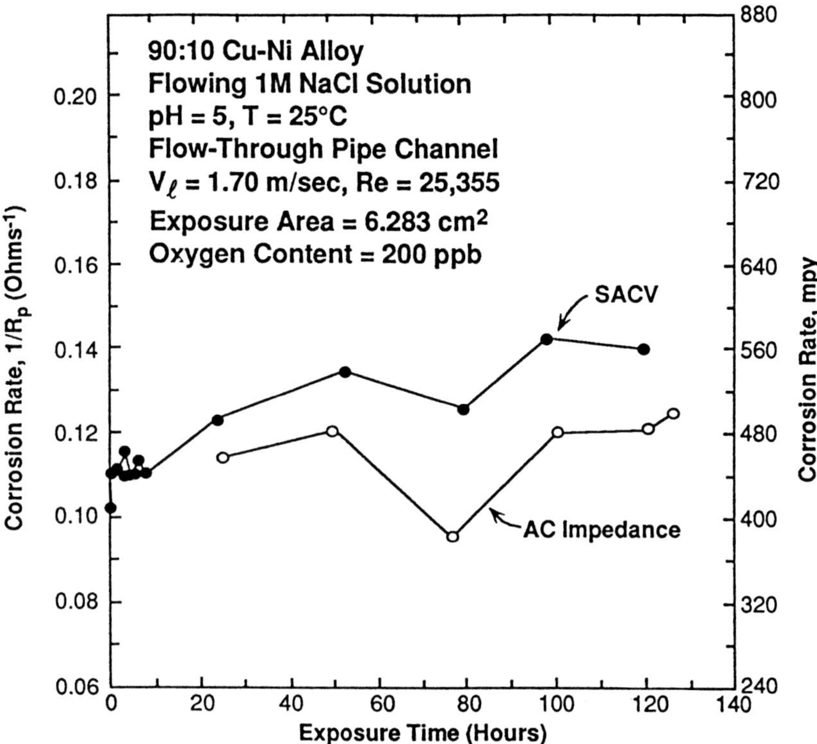

Figure 56   Variation of Corrosion Rate (as $1/R_p$) with Exposure Time for the Flow-Through Pipe Channel at Re = 25,355, pH = 5, T = 25°C (77°F)l [O_2] = 200 ppb

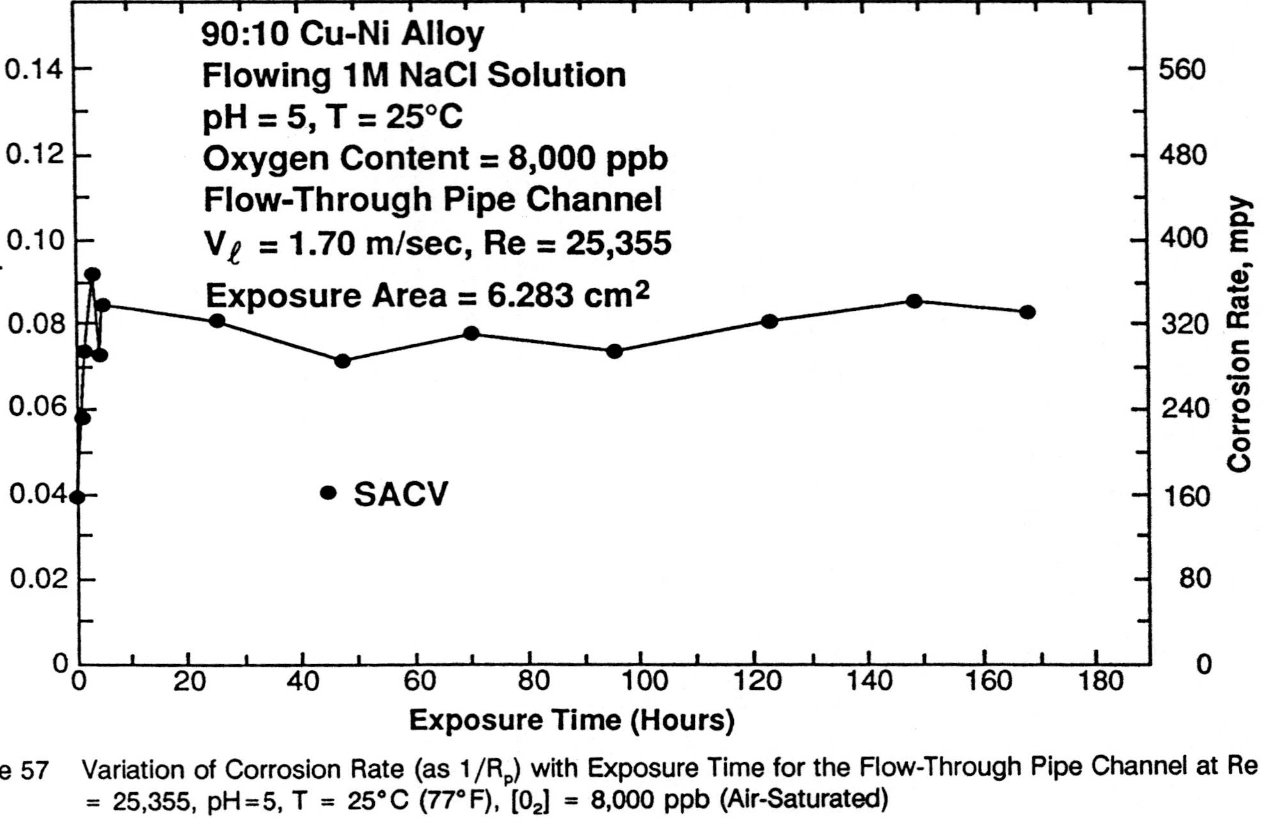

Figure 57   Variation of Corrosion Rate (as $1/R_p$) with Exposure Time for the Flow-Through Pipe Channel at Re = 25,355, pH=5, T = 25°C (77°F), [O_2] = 8,000 ppb (Air-Saturated)

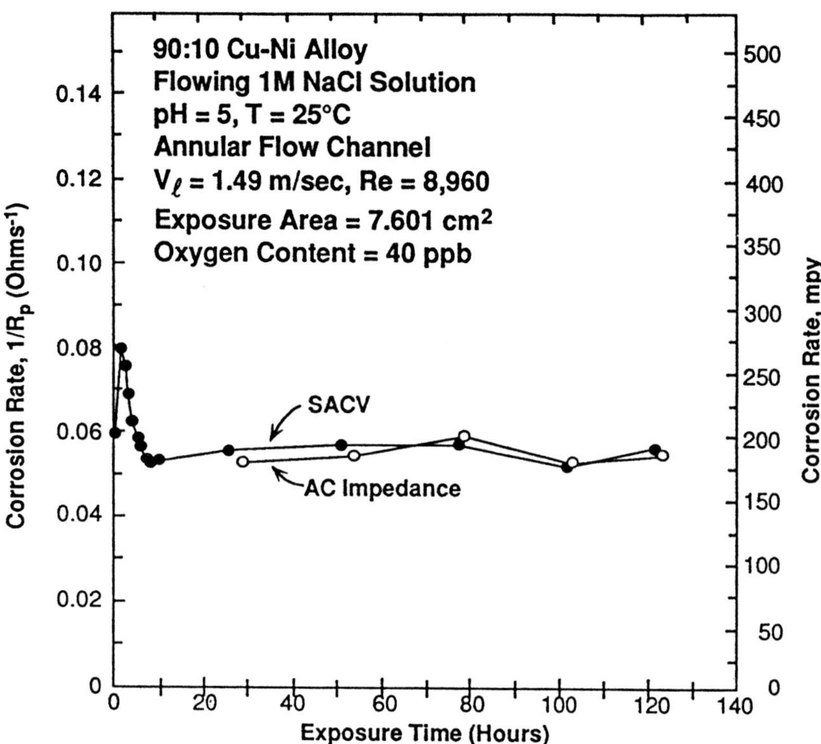

Figure 58   Variation of Corrosion Rate (as 1/R_p) with Exposure Time for the Annular Flow Channel at Re = 8,960, pH = 5, T = 25°C (77°F), [O_2] = 40 ppb

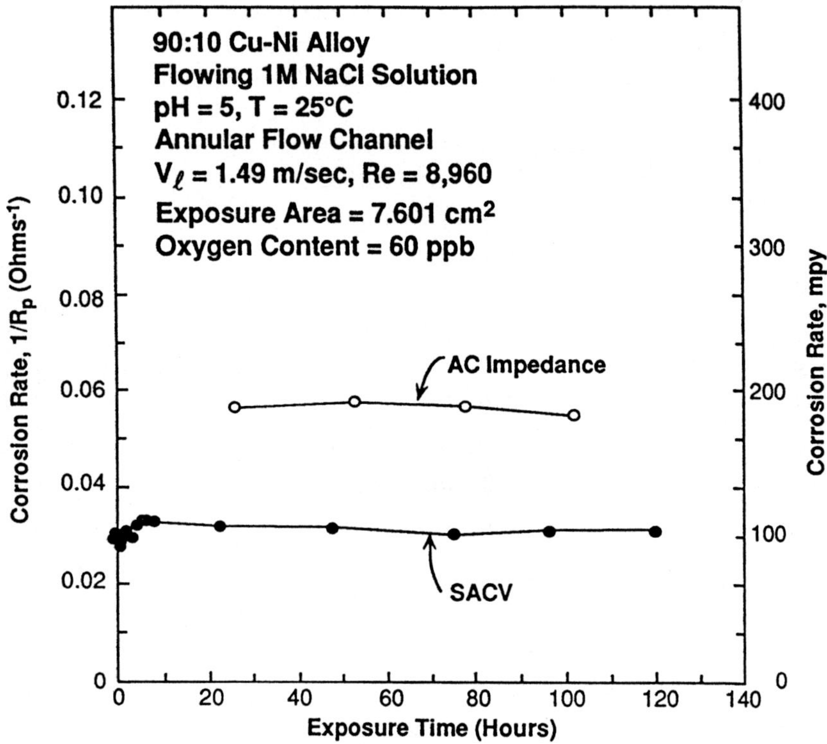

Figure 59   Variation of Corrosion Rate (as 1/R_p) with Exposure Time for the Annular Flow Channel at Re = 8,960, pH = 5, T = 25°C (77°F), [O_2] = 60 ppb

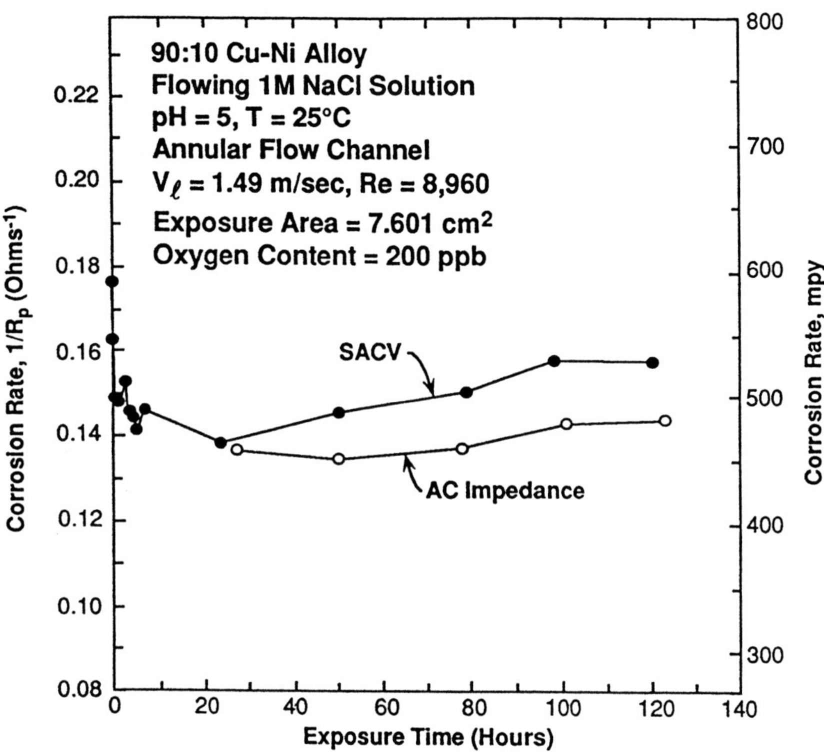

Figure 60   Variation of Corrosion Rate (as $1/R_p$) with Exposure Time for the Annular Flow Channel at Re = 8,960, pH = 5, T = 25°C (77°F), $[O_2]$ = 200 ppb

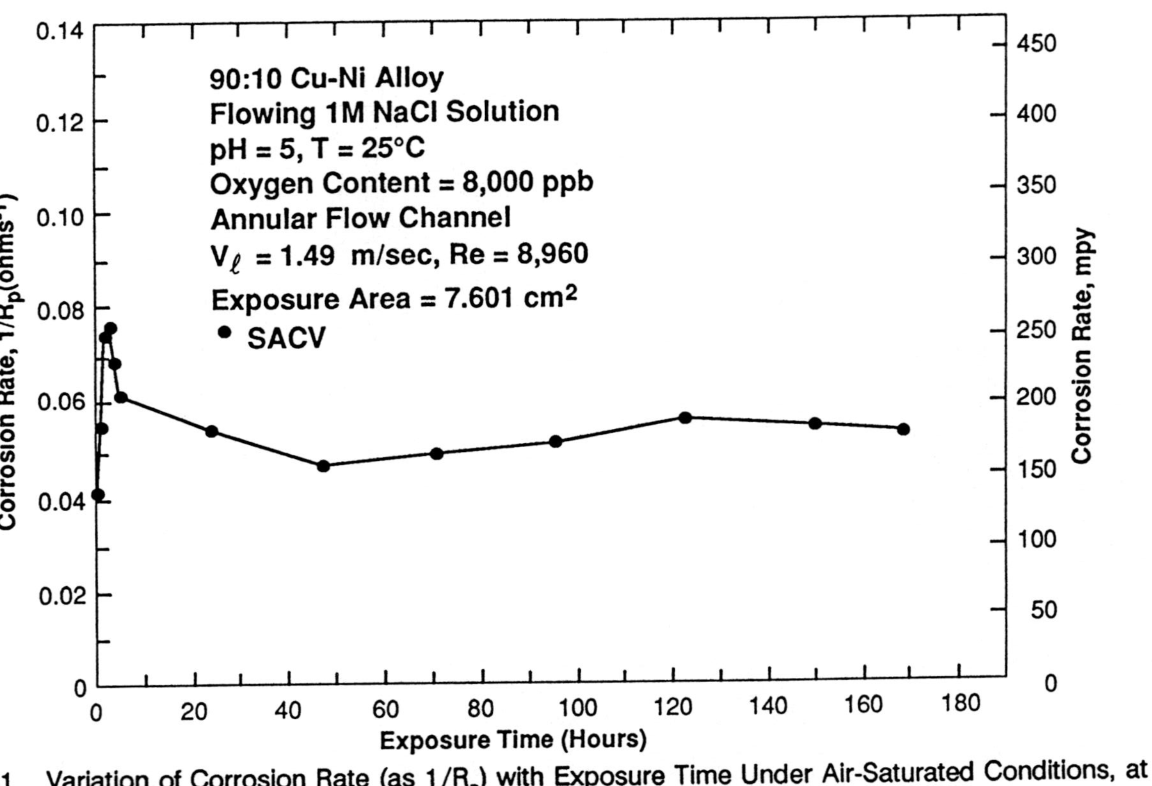

Figure 61   Variation of Corrosion Rate (as $1/R_p$) with Exposure Time Under Air-Saturated Conditions, at pH = 5, T = 25°C (77°F), for the Annular Flow Channel at Re = 8,960

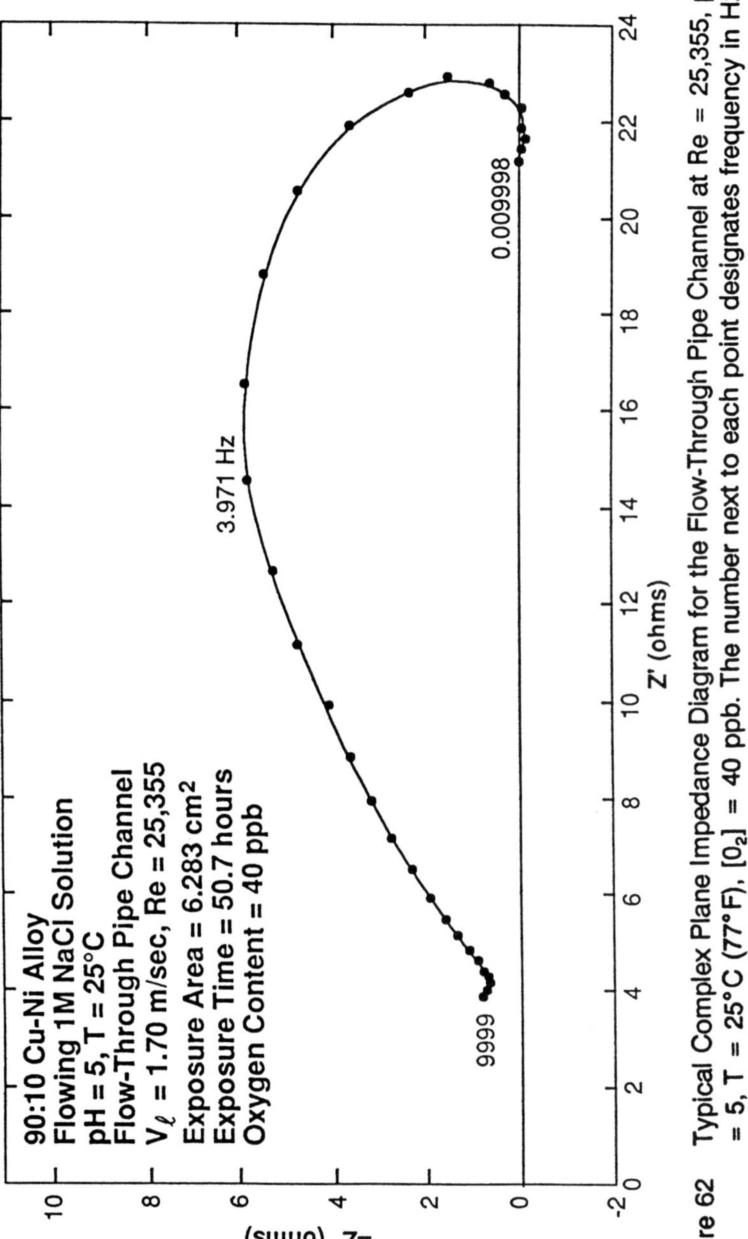

Figure 62   Typical Complex Plane Impedance Diagram for the Flow-Through Pipe Channel at Re = 25,355, pH = 5, T = 25°C (77°F), $[O_2]$ = 40 ppb. The number next to each point designates frequency in Hz.

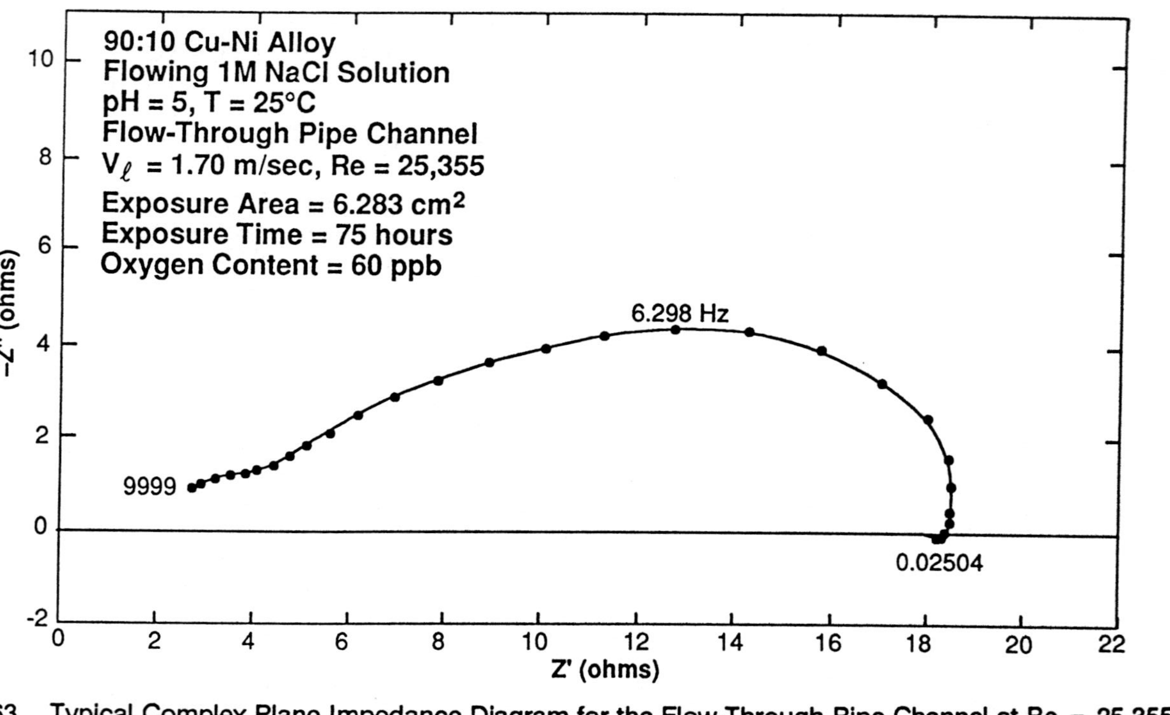

Figure 63   Typical Complex Plane Impedance Diagram for the Flow-Through Pipe Channel at Re = 25,355, pH = 5, T = 25°C (77°F), [O$_2$] = 60 ppb. The number next to each point designates frequency in Hz

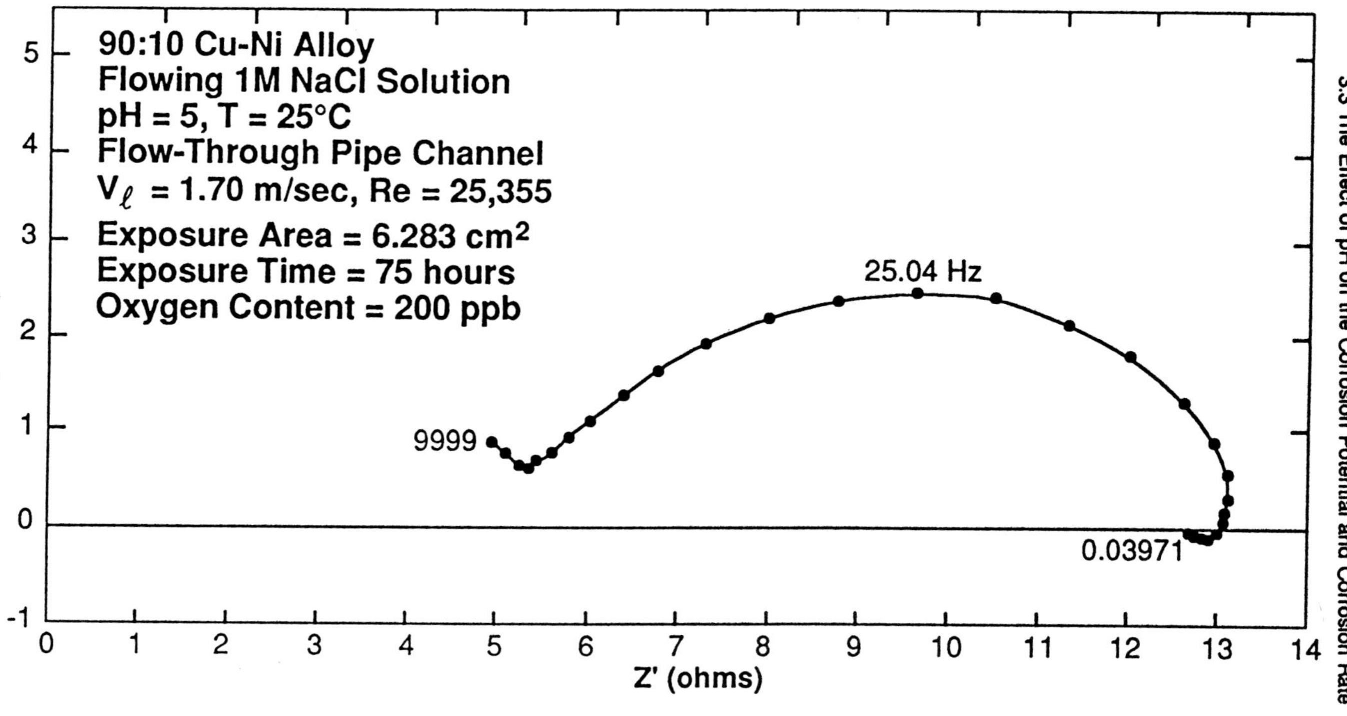

Figure 64   Typical Complex Plane Impedance Diagram for the Flow-Through Pipe Channel at Re = 25,355, pH = 5, T = 25°C (77°F), $[O_2]$ = 200 ppb. The number next to each point designates frequency in Hz

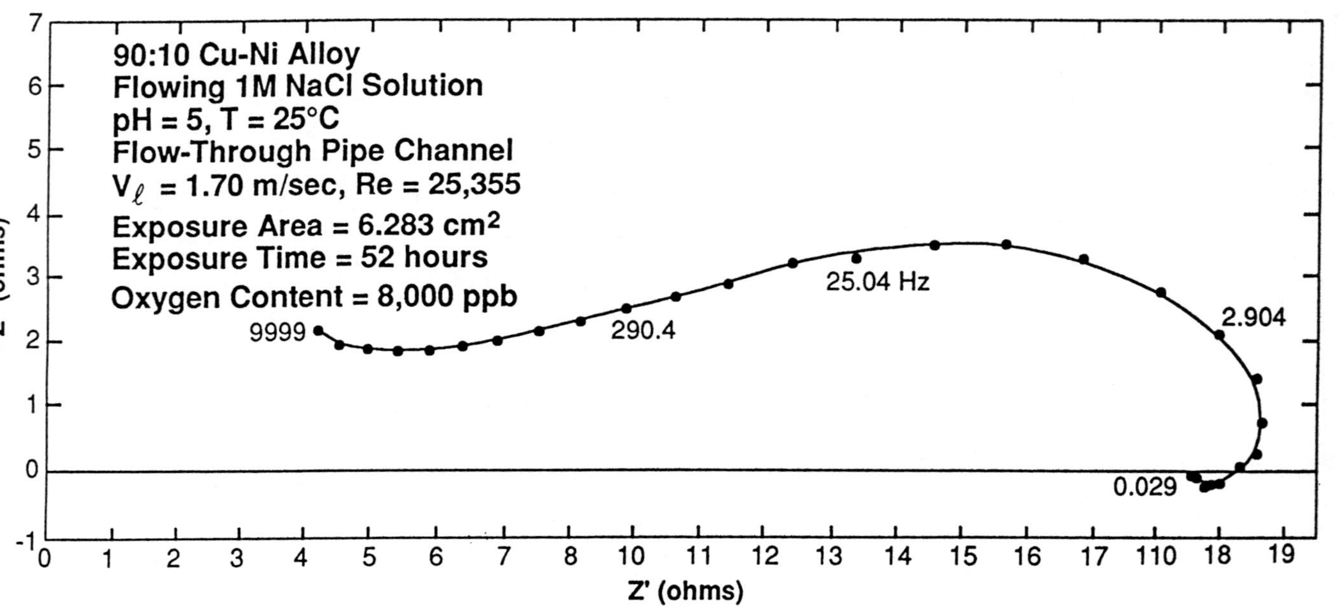

Figure 65   Typical Complex Plane Impedance Diagram for the Flow-Through Pipe Channel at Re = 25,355, pH = 5, T = 25°C (77°F), [O$_2$] = 8,000 ppb (Air-Saturated). The number next to each point designates frequency in Hz.

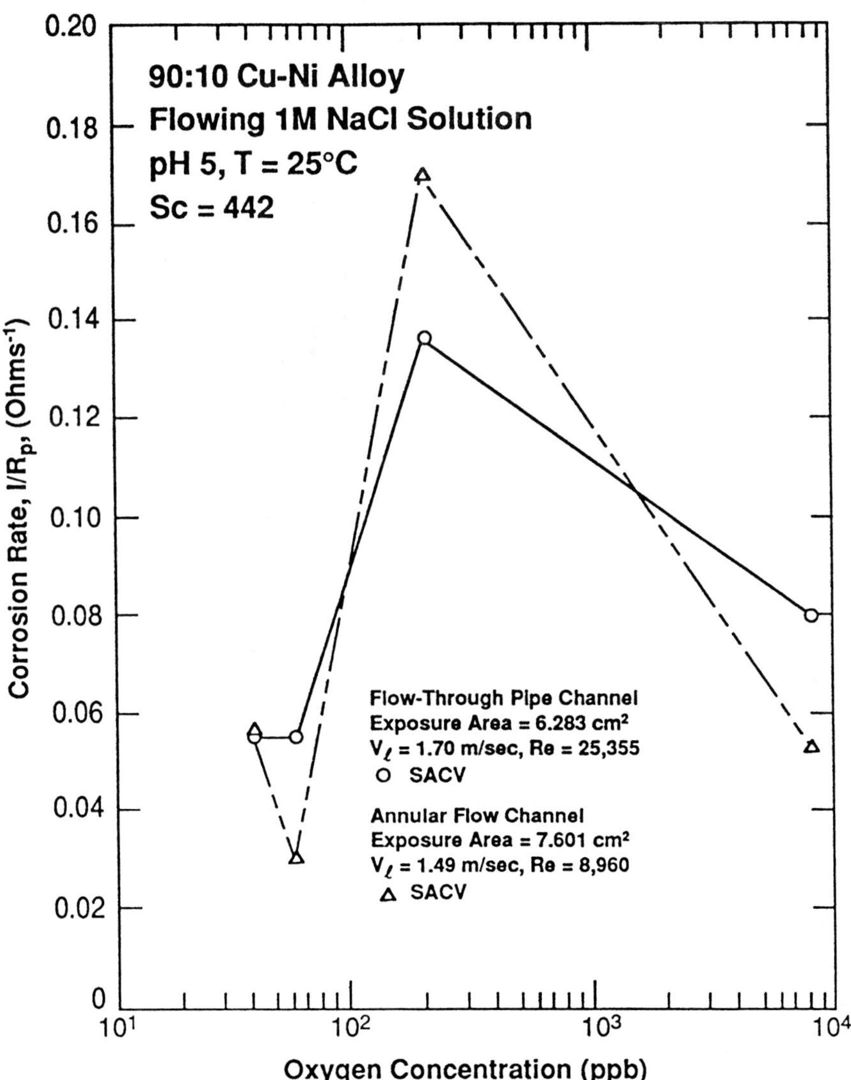

Figure 70   Variation of Corrosion Rate (as $1/R_p$) with Dissolved Oxygen Concentration for the Flow-Through Pipe Channel and the Annular Flow Channel

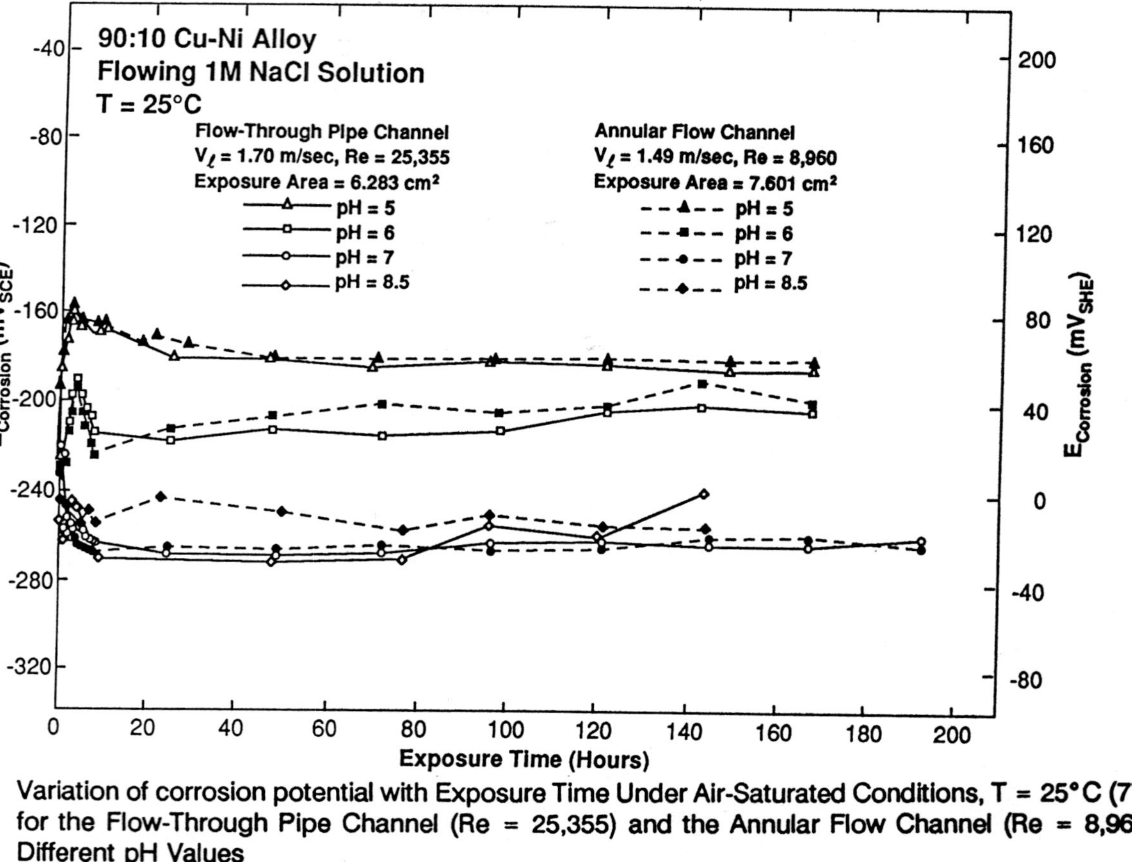

Figure 71   Variation of corrosion potential with Exposure Time Under Air-Saturated Conditions, T = 25°C (77°F), for the Flow-Through Pipe Channel (Re = 25,355) and the Annular Flow Channel (Re = 8,960) at Different pH Values

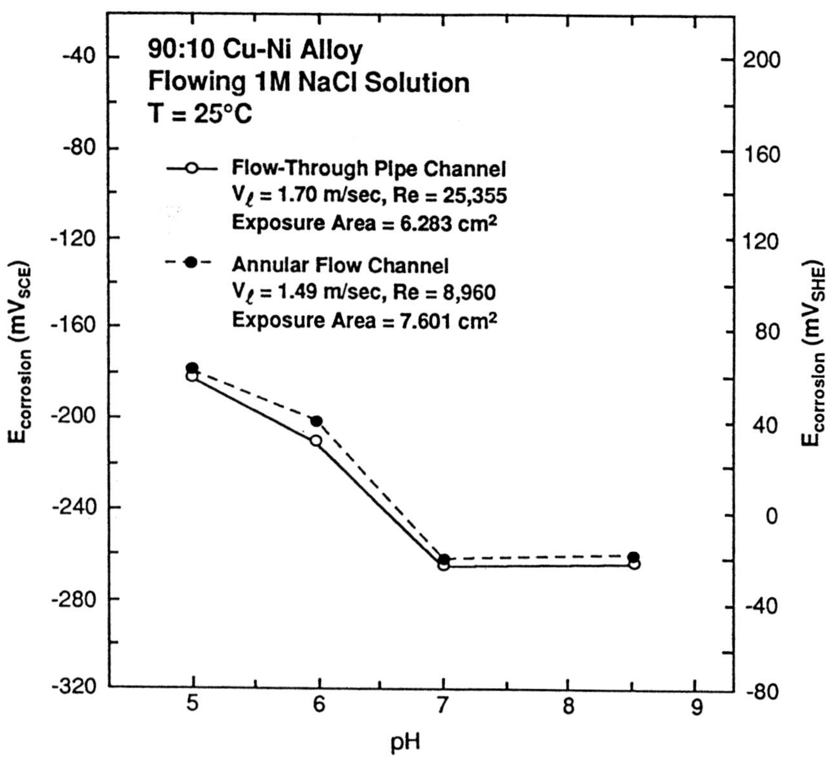

Figure 72   Variation of Corrosion Potential with pH

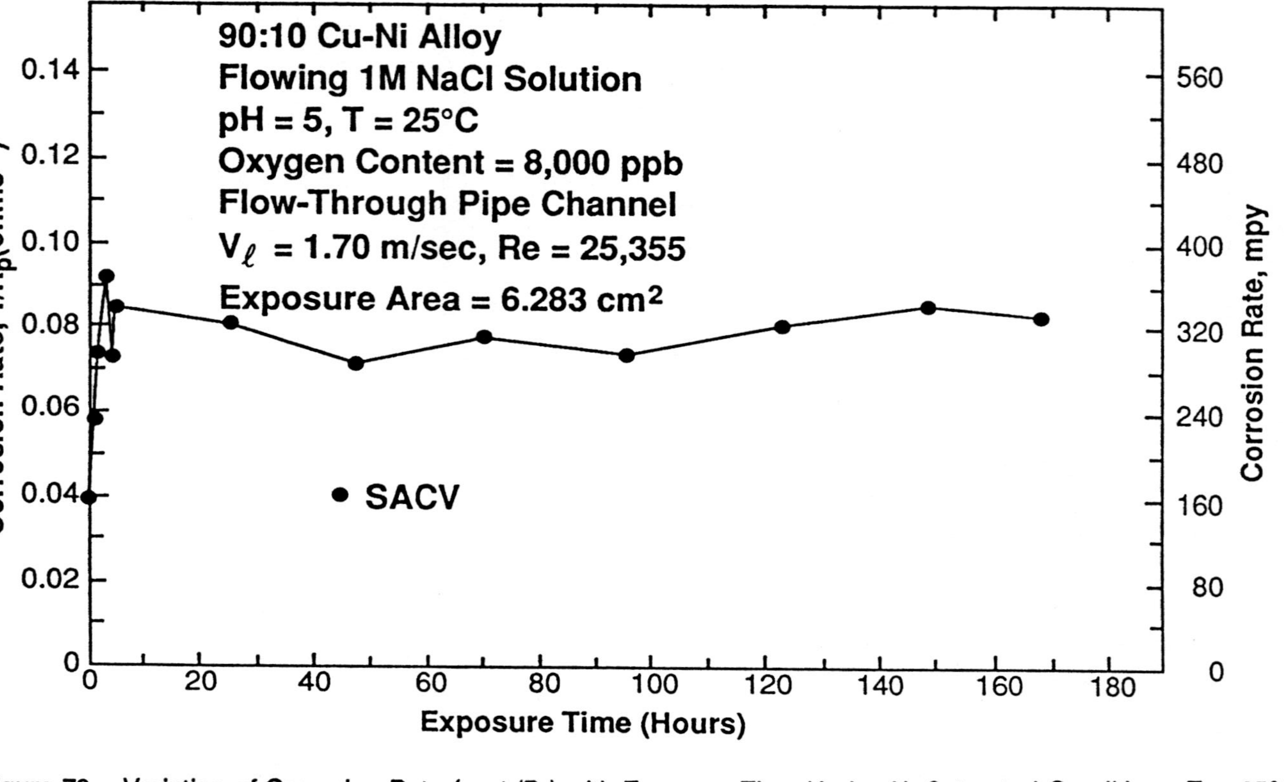

Figure 73   Variation of Corrosion Rate (as 1/R$_p$) with Exposure Time Under Air-Saturated Conditions, T = 25°C (77°F), Re = 25,355, for the Flow-Through Pipe Channel at pH = 5

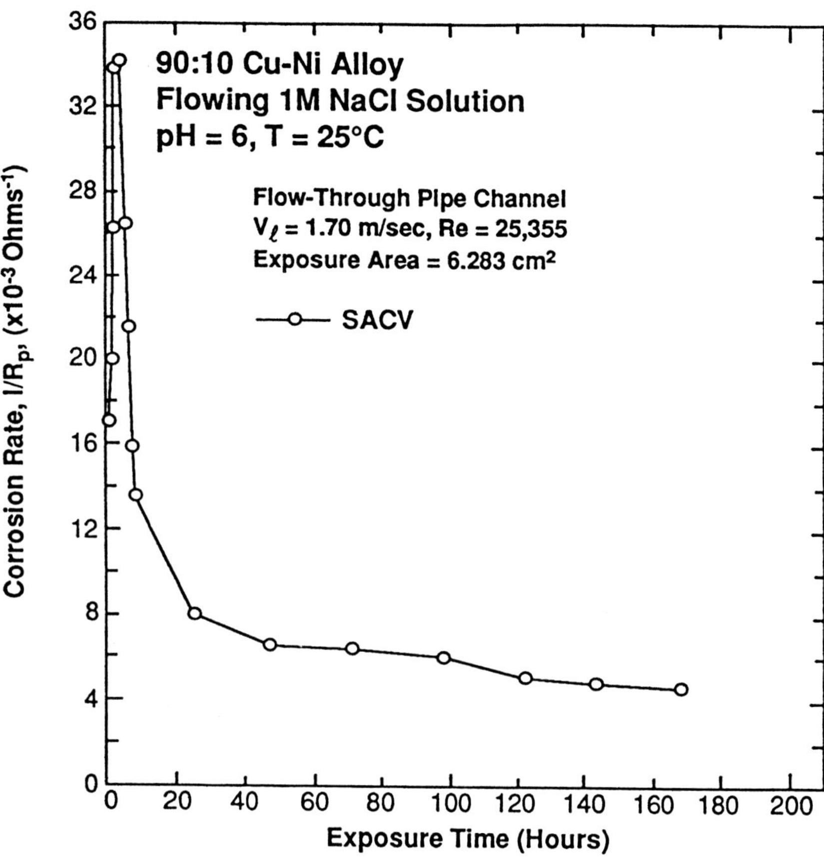

Figure 74   Variation of Corrosion Rate (as $1/R_p$) with Exposure Time Under Air-Saturated Conditions, T = 25°C (77°F), Re = 25,355, for the Flow-Through Pipe Channel at pH = 6

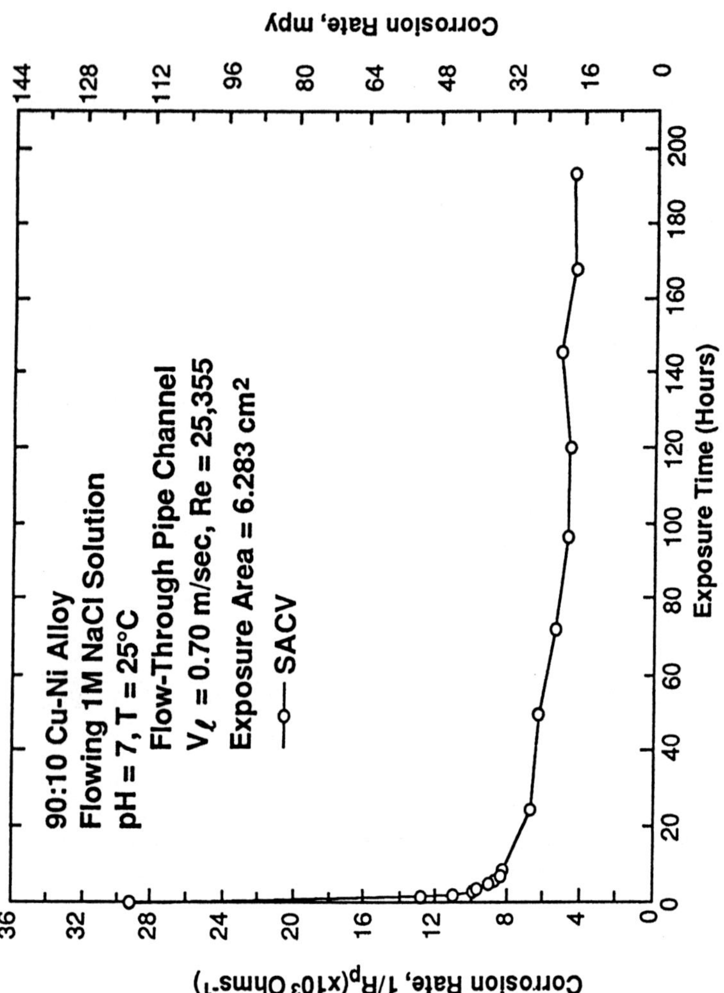

Figure 75   Variation of Corrosion Rate (as 1/R_p) with Exposure Time Under Air-Saturated Conditions, T = 25°C (77°F), Re = 25,355, for the Flow-Through Pipe Channel at pH = 7

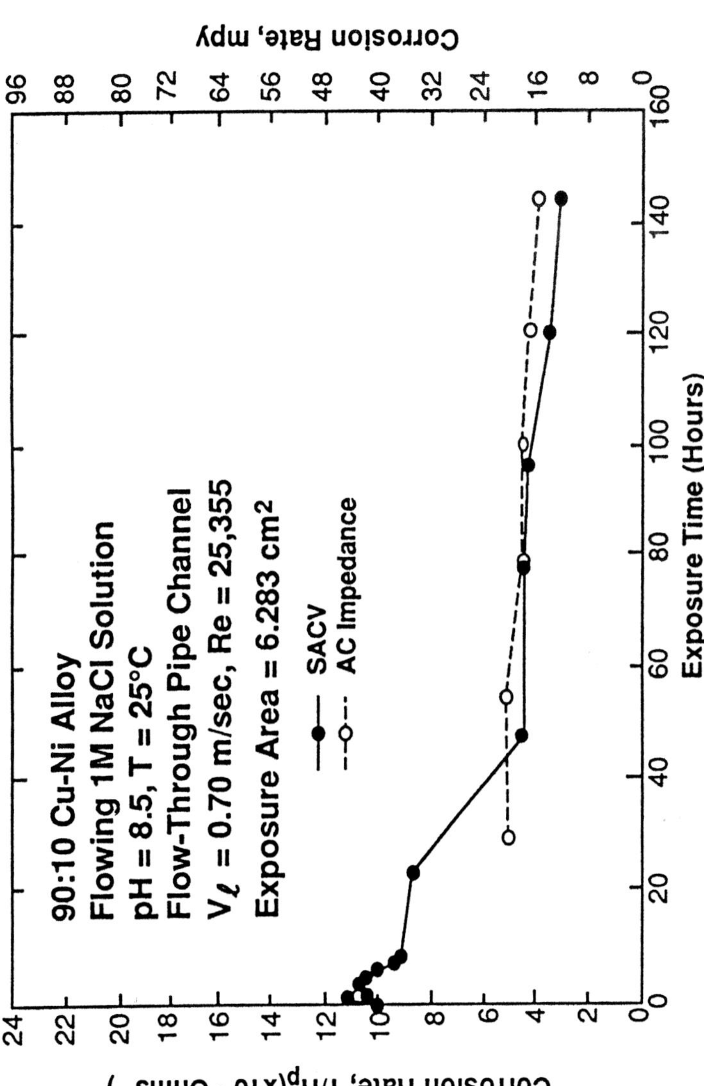

Figure 76   Variation of Corrosion Rate (as $1/R_p$) with Exposure Time Under Air-Saturated Conditions, T = 25°C (77°F), Re = 25,355, for the Flow-Through Pipe Channel at pH = 8.5

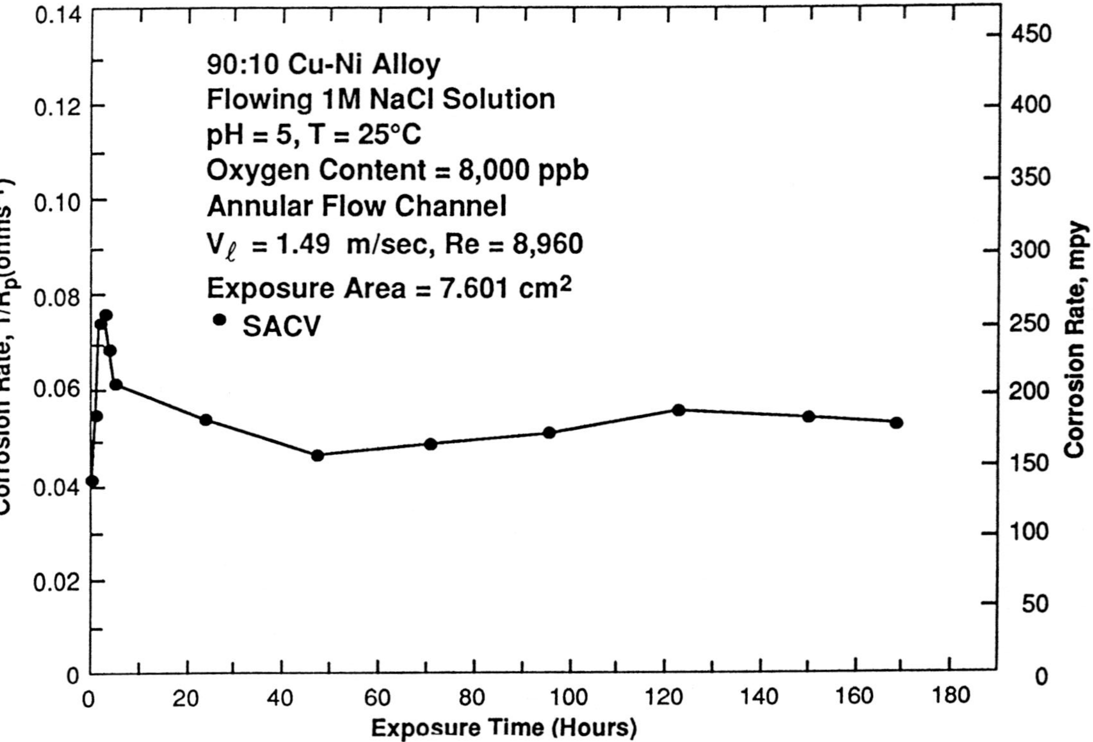

Figure 77   Variation of Corrosion Rate (as 1/R$_p$) with Exposure Time Under Air-Saturated Conditions, T = 25°C (77°F), Re = 8,960, for the Annular Flow Channel at pH = 5

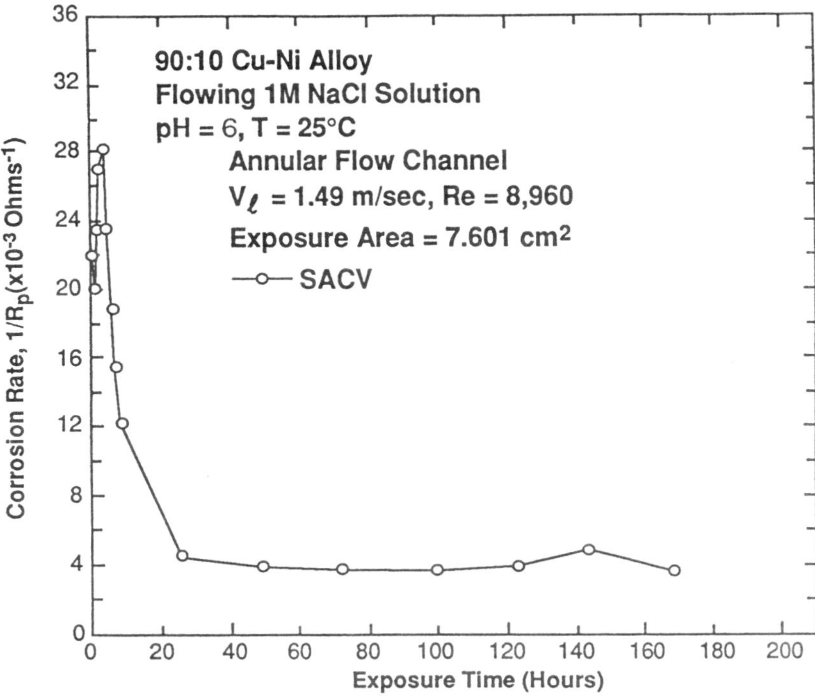

Figure 78   Variation of Corrosion Rate (as $1/R_p$) with Exposure Time Under Air-Saturated Conditions, T = 25°C (77°F), Re = 8,960, for the Annular Flow Channel at pH = 6

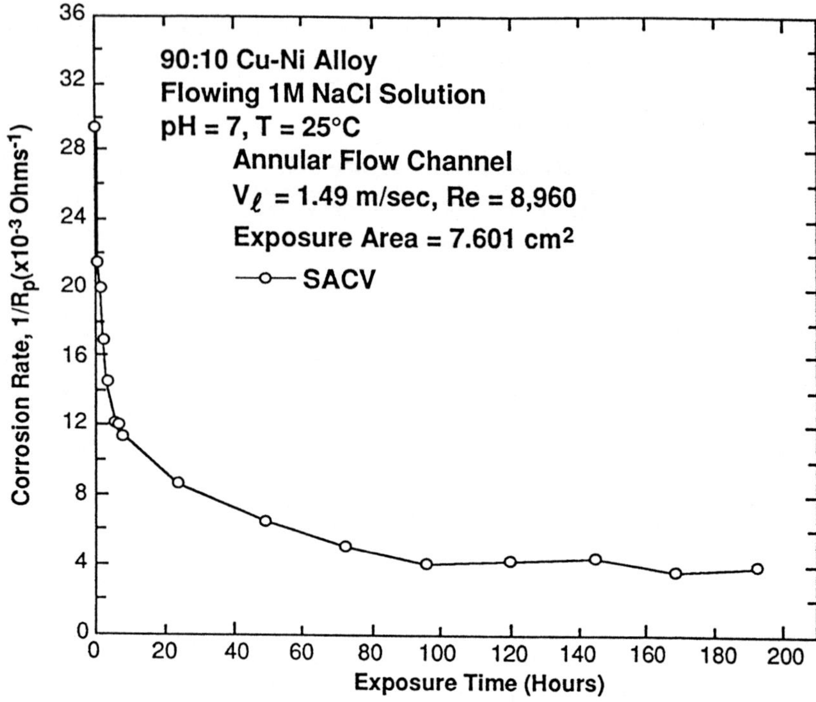

Figure 79   Variation of Corrosion Rate (as $1/R_p$) with Exposure Time Under Air-Saturated Conditions, T = 25°C (77°F), Re = 8,960, for the Annular Flow Channel at pH = 7

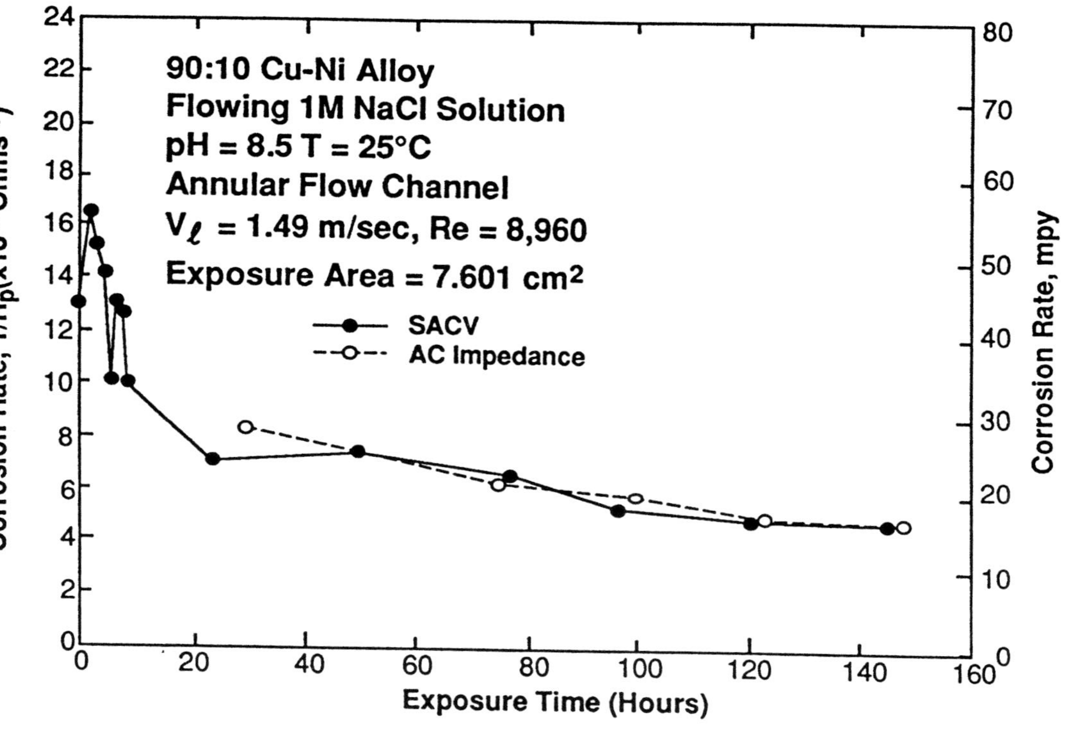

Figure 80   Variation of Corrosion Rate (as $1/R_p$) with Exposure Time Under Air-Saturated Conditions, T = 25°C (77°F), Re = 8,960, for the Annular Flow Channel at pH = 8.5

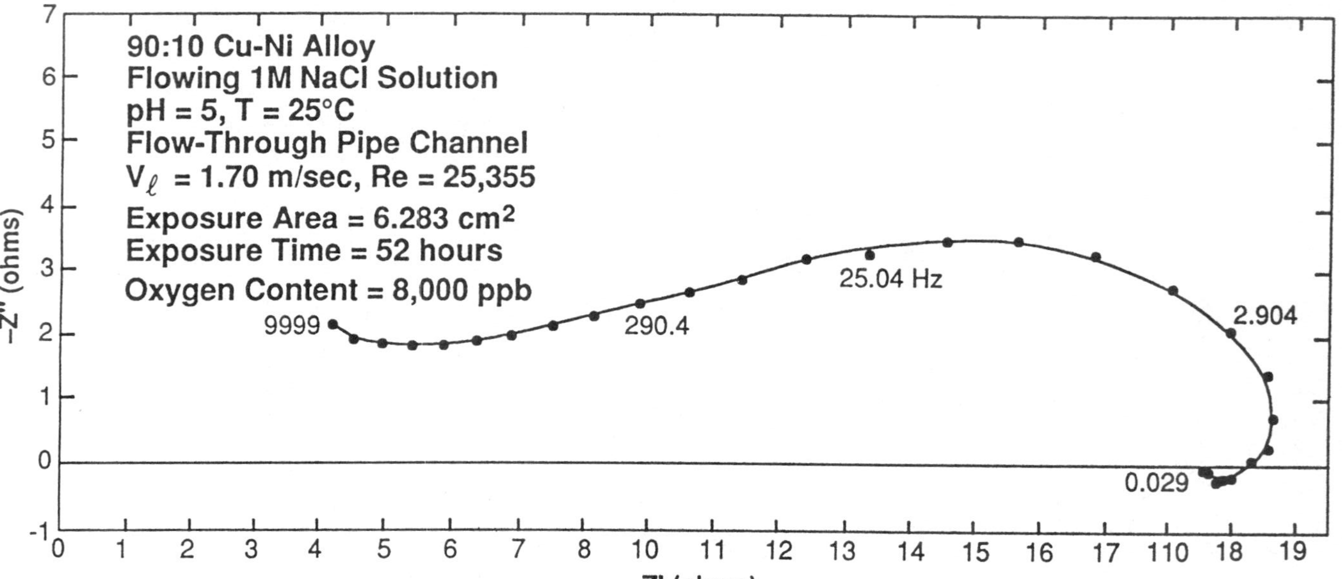

Figure 81   Typical Complex Plane Impedance Diagram for the Flow-Through Pipe Channel Under Air-Saturated Conditions, T = 25°C (77°F), Re = 25,355, at pH = 5. The number next to each point designates frequency in Hz.

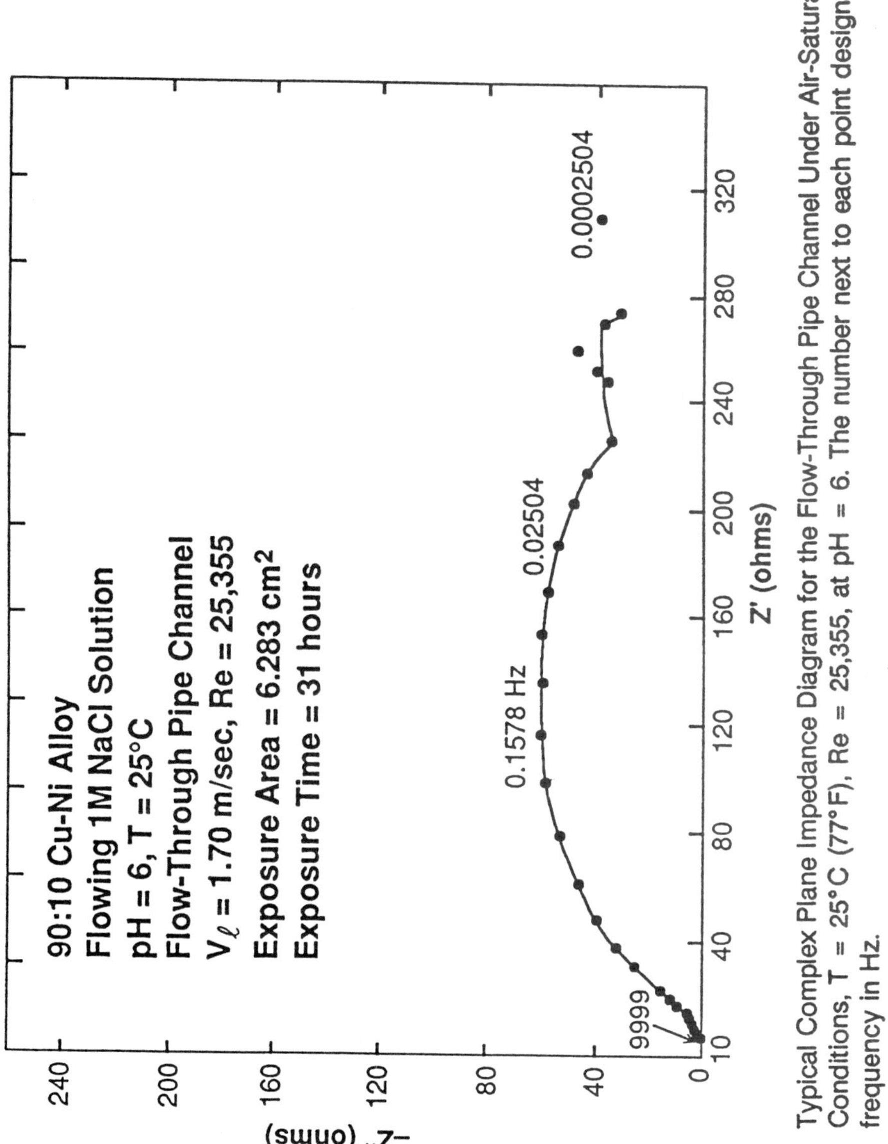

Figure 82   Typical Complex Plane Impedance Diagram for the Flow-Through Pipe Channel Under Air-Saturated Conditions, T = 25°C (77°F), Re = 25,355, at pH = 6. The number next to each point designates frequency in Hz.

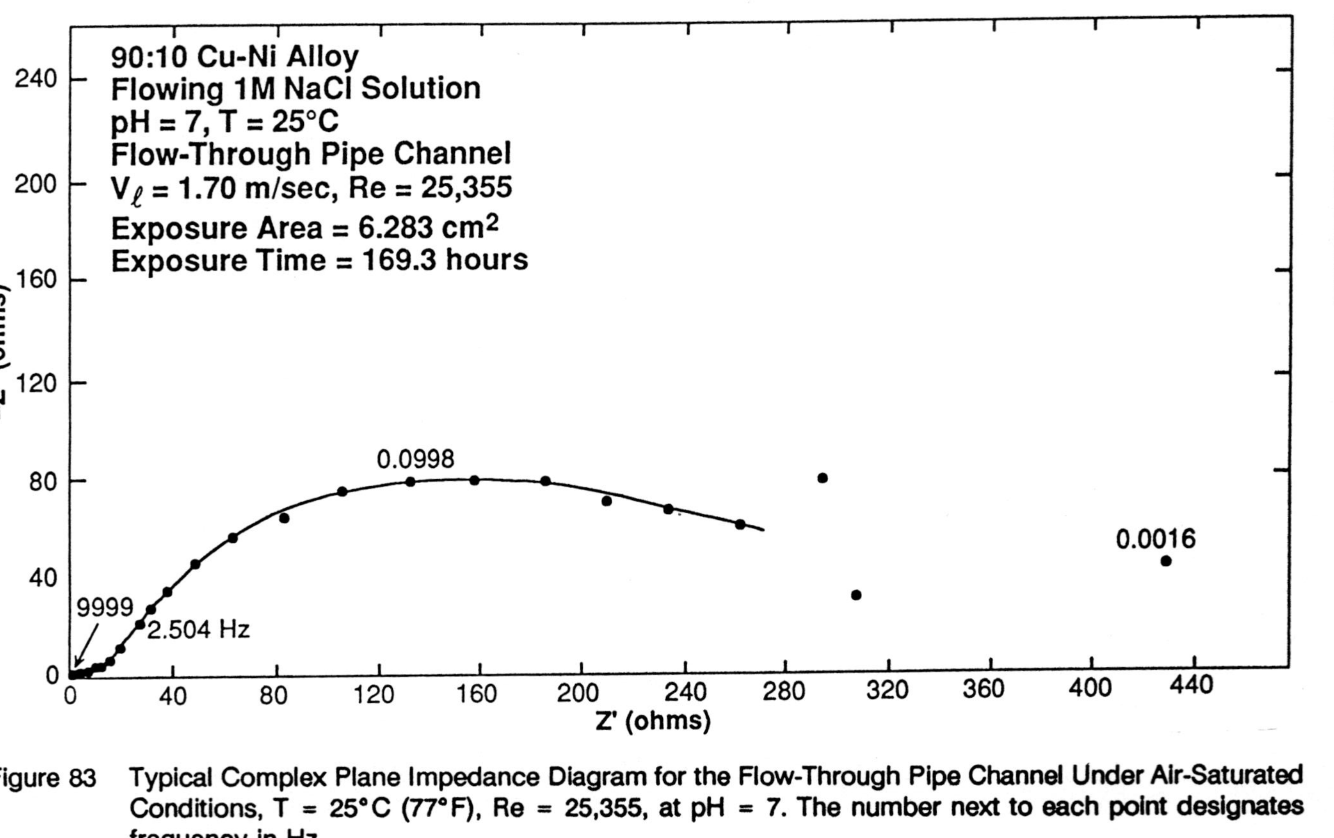

Figure 83   Typical Complex Plane Impedance Diagram for the Flow-Through Pipe Channel Under Air-Saturated Conditions, T = 25°C (77°F), Re = 25,355, at pH = 7. The number next to each point designates frequency in Hz.

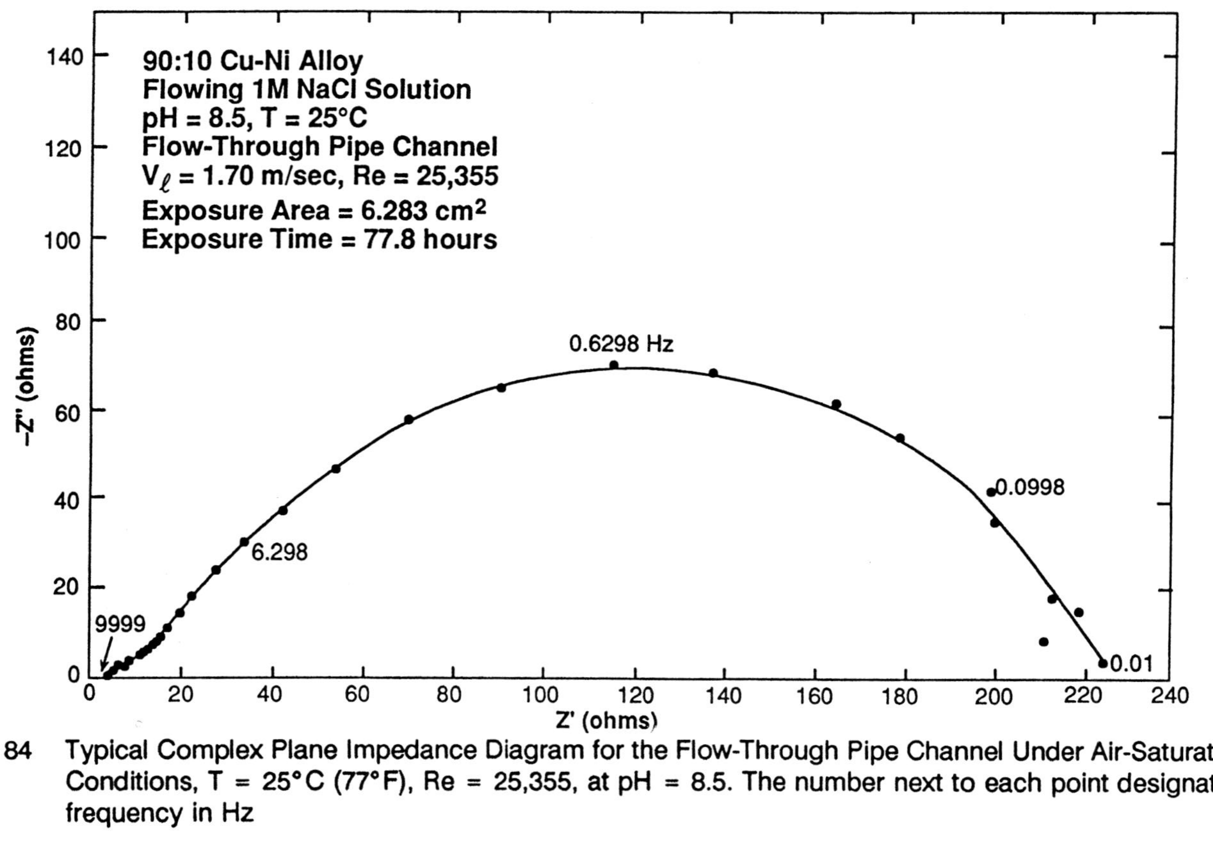

Figure 84   Typical Complex Plane Impedance Diagram for the Flow-Through Pipe Channel Under Air-Saturated Conditions, T = 25°C (77°F), Re = 25,355, at pH = 8.5. The number next to each point designates frequency in Hz

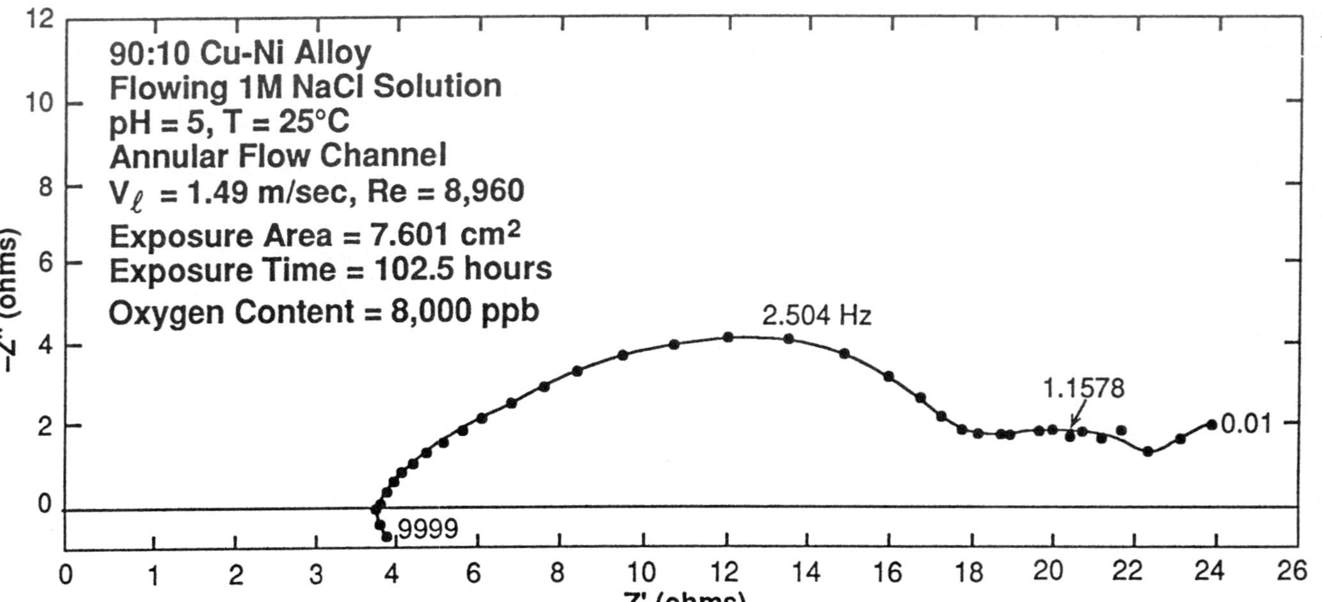

Figure 85   Typical Complex Plane Impedance Diagram for the Annular Flow Channel Under Air-Saturated Conditions, T = 25°C (77°F), Re = 8,960, at pH = 5. The number next to each point designates frequency in Hz.

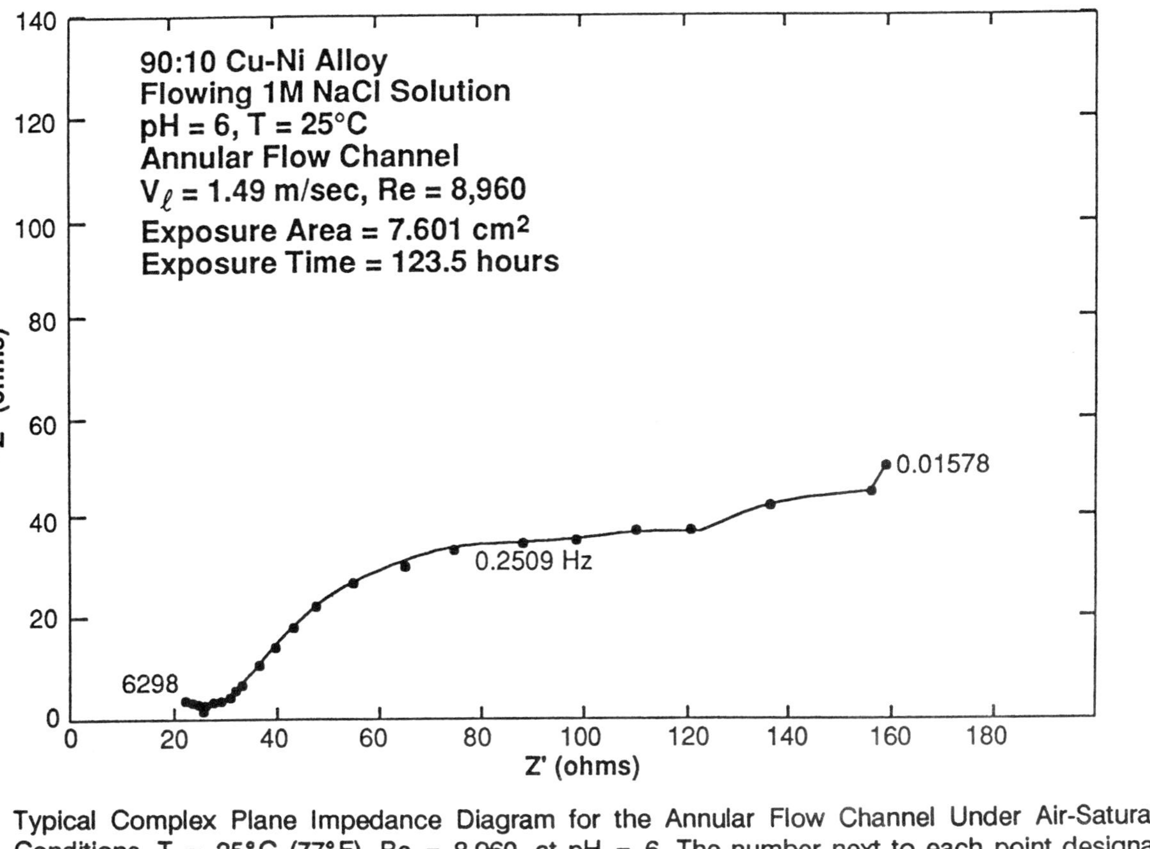

Figure 86   Typical Complex Plane Impedance Diagram for the Annular Flow Channel Under Air-Saturated Conditions, T = 25°C (77°F), Re = 8,960, at pH = 6. The number next to each point designates frequency in Hz.

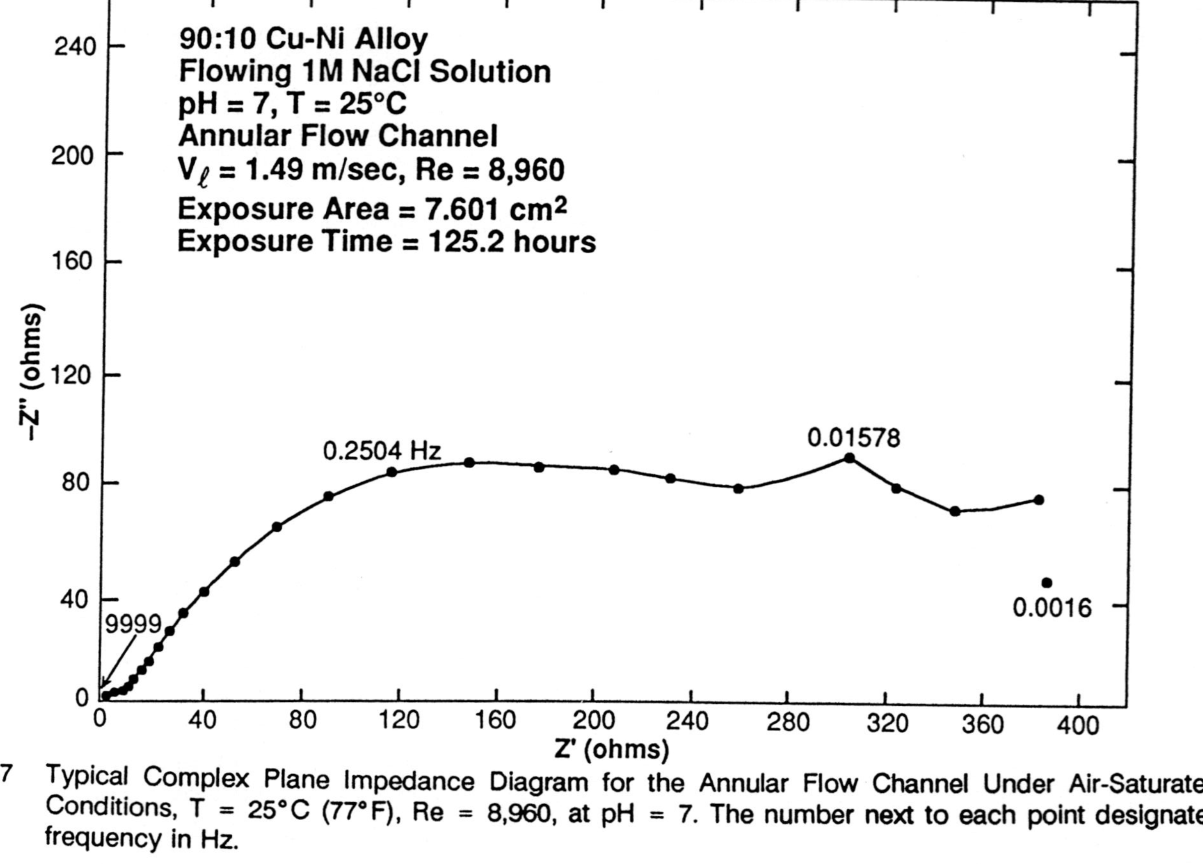

Figure 87   Typical Complex Plane Impedance Diagram for the Annular Flow Channel Under Air-Saturated Conditions, T = 25°C (77°F), Re = 8,960, at pH = 7. The number next to each point designates frequency in Hz.

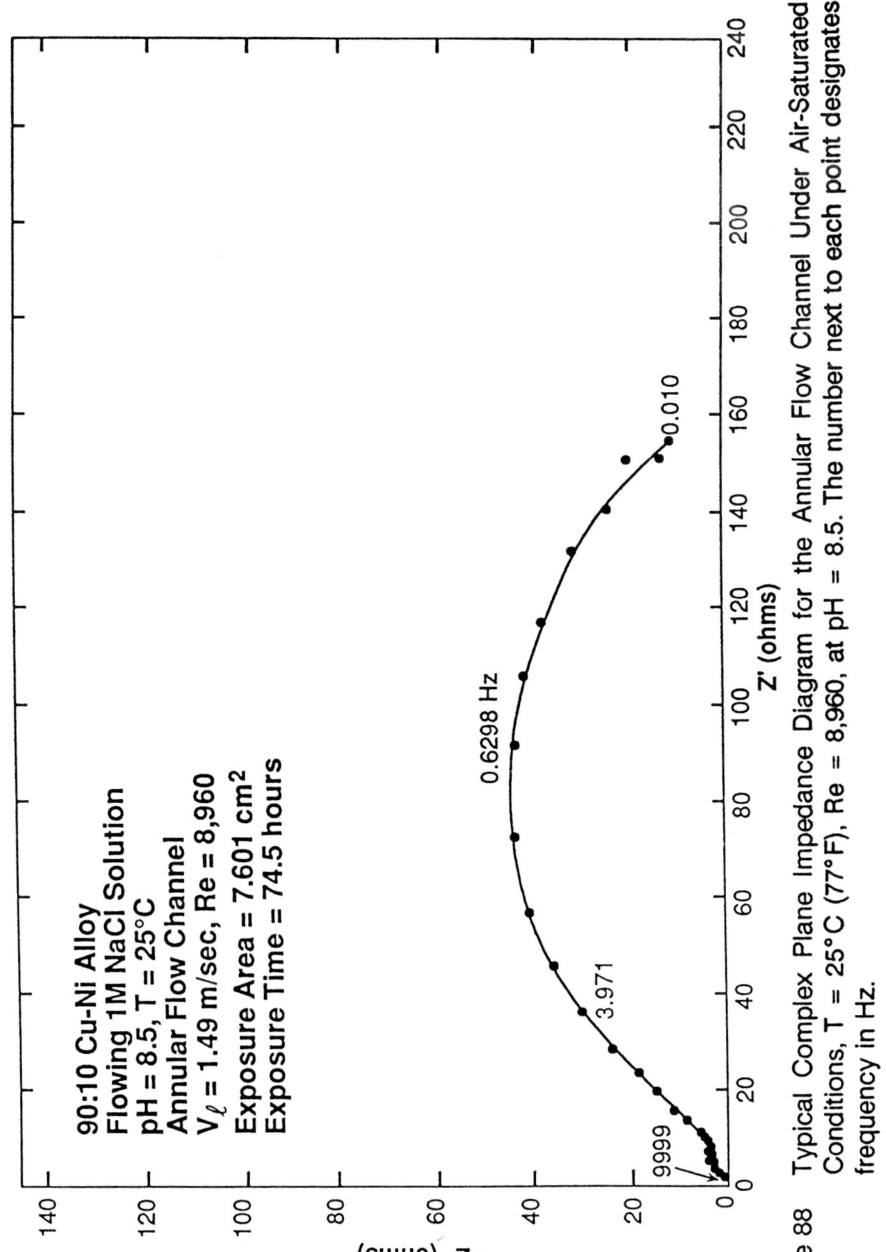

Figure 88   Typical Complex Plane Impedance Diagram for the Annular Flow Channel Under Air-Saturated Conditions, T = 25°C (77°F), Re = 8,960, at pH = 8.5. The number next to each point designates frequency in Hz.

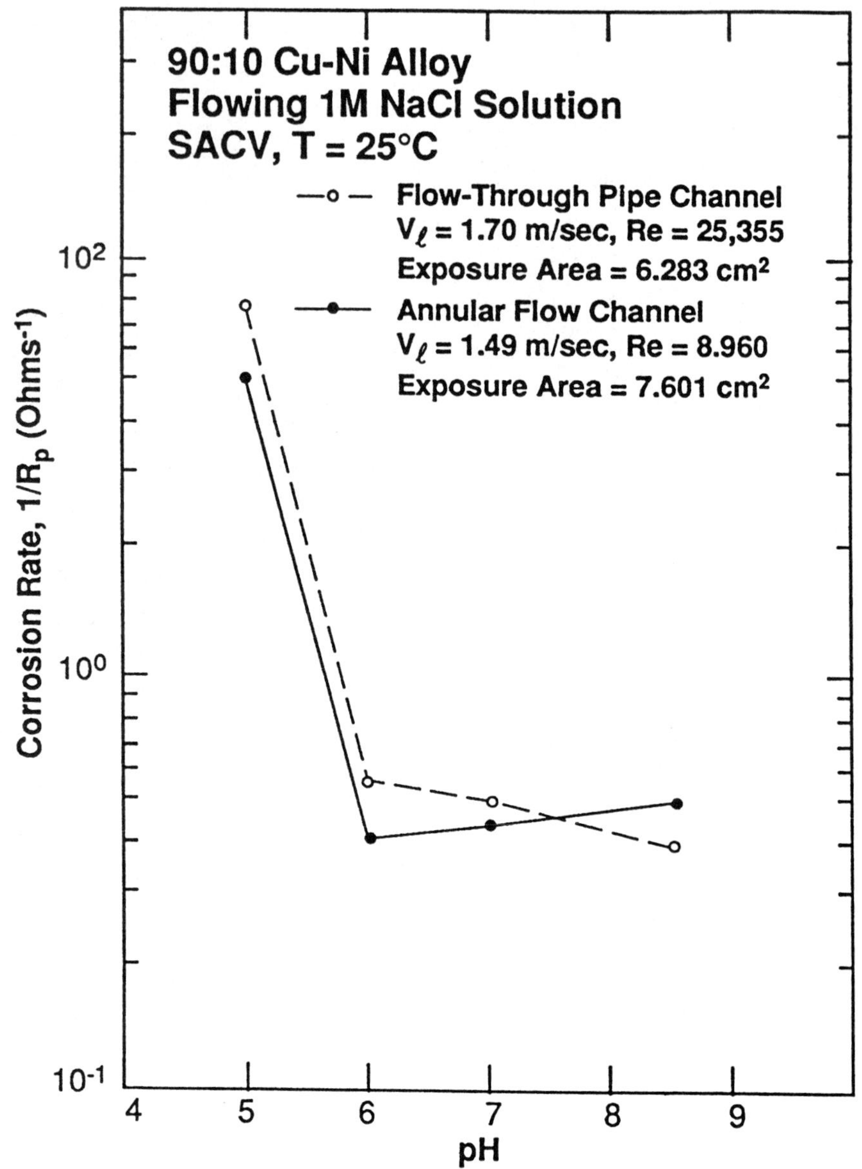

Figure 89   Variation of Corrosion Rate (as $1/R_p$) with pH

CHAPTER
FOUR

TECHNOLOGICAL IMPLICATIONS

One of the goals of this study was to develop inexpensive test systems for evaluating the corrosion of materials under carefully controlled hydrodynamic conditions. This objective was undertaken so that materials degradation phenomena could be studied in the laboratory under conditions that effectively simulate those found in the field. Furthermore, we sought to develop corrosion rate/hydrodynamic correlations that would permit transfer of the laboratory data to field situations, provided that the hydrodynamic characteristics of the field system were known. Both objectives have been achieved, although the latter has been tested only for three rather restricted geometries.

The three controlled hydrodynamic systems developed in this work (flow-through pipe, annular flow channel, and rotating cylinder) have been shown to be effective test systems for the evaluation of corrosion phenomena under carefully controlled hydrodynamic conditions. However, because of its compactness and simplicity, we believe that the rotating cylinder system is the most effective and convenient test method for use in the laboratory, and we recommend that this geometry be adopted for routine corrosion testing and materials evaluation activities. The work reported in this study provides the important link between the hydrodynamic and corrosion characteristics of the rotating cylinder system and geometries that are frequently encountered in the field, most notably circular pipes.

Also of technological significance is the work reported on the use of electrochemical techniques for corrosion testing. This study has evaluated both small amplitude cyclic voltammetry (SACV) and alternating current (AC) impedance techniques, on a competitive basis, for the *in situ* measurement of polarization resistance and has compared the corrosion rates calculated from these electrochemical data with directly measured weight losses. In general, satisfactory agreement was obtained, particularly considering the fact that very high corrosion rates were observed at short times, and hence, considerable error could be introduced in integrating the polarization resistance data over the exposure time. The problems associated with the electrochemical evaluation of corrosion rates for Cu-Ni alloys have been discussed in detail by Syrett and Macdonald,[39] and the errors induced by high-corrosion rates at short times are frequently much larger than the differences observed in this work between the calculated corrosion rates and the measured weight losses. It is emphasized that these differences do not imply that the electrochemical techniques are incapable of correctly measuring corrosion rates, only that it is necessary to collect polarization resistance data at short times.

Finally, comparison of the performances of the AC impedance and SACV techniques for monitoring corrosion rates *in situ* has shown that the latter provides more consistent and accurate values for the polarization resistance. Theoretically, both small perturbation techniques should yield identical data. However, the SACV technique has proved to be superior to the AC impedance technique for the measurement of polarization resistance, and it is recommended that this technique be used in the in-plant environment, at least for monitoring the rate of corrosion of Cu-Ni alloys in sodium chloride containing solutions.

FIVE

SUMMARY AND CONCLUSIONS

5.1 SUMMARY AND CONCLUSIONS

- For all the hydrodynamic systems, at each Reynolds number (Re) studied, the corrosion potential as well as the rate of corrosion increased with exposure time in the first few hours of exposure but subsequently decreased to a nearly constant value for long exposure times. The average steady state values of corrosion potential and corrosion rate increased with increasing Re number.

- Good agreement was obtained between the small amplitude cyclic voltammetry (SACV) and alternating current (AC) impedance techniques for measuring polarization resistance, and the calculated weight losses were in reasonable accord with directly measured values.

- The mass transfer rate correlations for the hydrodynamic systems studied are in general agreement with published data.

- The corrosion rate measured in all the hydrodynamic systems increased linearly with increasing Sherwood numbers (Sh_1 and Sh_2) and mass transfer coefficients (K_1 and K_2) on a log-log scale. This shows that the rate of corrosion is controlled by the rate of mass transfer.

- In the region of high Re numbers (Re > Re$_{critical}$ ≈ 10,000), transfer of the corrosion rate from one geometrical system to another can be made based on the mass transfer coefficient for the diffusion boundary layer, K_h, calculated from a previously established mass transfer rate equation. The transfer of the corrosion rate from one geometry to another can also be achieved based on the overall mass transfer coefficient K_1; however, the relation between the mass transfer coefficient in the surface film, K_f, and the Re number for each hydrodynamic system must be experimentally determined.

5.2 THE EFFECT OF DISSOLVED OXYGEN CONCENTRATION ON THE CORROSION RATE

- The corrosion rate and the corrosion potential measured in the flow-through pipe channel and the annular flow channel first increased but then decreased with increasing dissolved oxygen concentration.

- Good agreement was obtained between the SACV and AC impedance techniques for the measurement of polarization resistance, and the calculated weight losses were in reasonable accord with the values directly measured.

5.3 THE EFFECT OF pH ON CORROSION RATE

- For the flow-through pipe channel and the annular flow channel, the average steady state corrosion potential first decreased with increasing pH but then became independent of it at pH > 7.

- The rate of corrosion measured in the flow-through pipe channel and the annular flow channel first decreased with increasing pH but then tended toward a constant value at pH > 6.

- Good agreement was obtained between the SACV and AC impedance techniques for the measurement of polarization resistance at pH = 8.5, and the calculated weight losses were in reasonable accord with directly measured values.

REFERENCES

1. T.K. Ross, B.P.L. Hitchen, *Corrosion Science* 1, 65(1961).

2. E. Heitz, *Werkst. U. Korr.* 15, 63(1964).

3. I. Cornet, E.A. Barrington, G.U. Behrsing, *J. Electrochem. Soc.* 108, 947(1961).

4. D.J. Pickett, *Electrochemical Reactor Design* (New York, NY: Elsevier Science Publishers, 1977).

5. R.B. Bird, W.E. Steward, E.N. Lightfoot, *Transport Phenomena* (New York, NY: John Wiley and Sons, Inc., 1960).

6. J.R. Selman, C.W. Tobias, *Advances in Chemical Engineering*, Vol. 10, eds. T.B. Drew, G.R. Cokelet, T. Vermeulen (New York, NY: Academic Press, 1978), p.212.

7. J.R. Selman, C.W. Tobias, *Advances in Chemical Engineering*, Vol. 10 (New York, NY: Academic Press, 1978), p.212.

8. B. Poulson, *Corrosion Science* 23, 391(1983).

9. W.J. Blaedal, C.L. Olson, L.R. Sharma, *Anal. Chem.* 35, 2100(1963).

10. L.R. Sharma, J. Dutt, *Indian J. Chem.* 6, 593(1968).

11. M.A. Devanathan, V. Guruswamy, *Sov. Electrochem.* 7, 1946(1971). (In English)

12. T.H. Chilton, A.P. Colburn, *Industrial and Engineering Chemistry* 26, 1183(1934).

13. P. Harriot, R.M. Hamilton, *Chem. Eng. Sci.* 20, 1023(1965).

14. M. Coney, CERL Report RD/L/N197/80, Nov. 1980.

15. F.P. Berger, KFEL Hue, *Int. J. Heat and Mass Transfer* 20, 1185(1977).

16. C.S. Lin, E.B. Denton, H.S. Gaskell, G.L. Putman, *Ind. Eng. Chem.* 43, 2136(1951).

17. W.L. Friend, A.B. Metzner, *AIChE J.* 4, 393(1958).

18. T.K. Ross, A.A. Wragg, *Electrochim Acta* 10, 1093(1965).

19. C.C. Monard, J.F. Pelton, *Trans. Amr. Inst. Chem. Engrs.* 38, 593(1943).

20. W.H. Linton, T.K. Sherwood, *Chem. Eng. Prog.* 46, 258(1950).

21. I. Cornet, R. Kappesser, *Trans. Inst. Chem. Eng.* 47, 194(1969).

22. D.R. Gabe, *J. Applied Electrochemistry* 4, 91(1974).

23. V.G. Levich, *Physiochemical Hydrodynamics* (New York, NY: Prentice Hall, 1962).

24. R. Bird, W.E. Stewart, E.N. Lightfoot, *Transport Phenomena* (New York, NY: John Wiley and Sons, Inc., 1960).

25. A. Kimla, F. Strafeld, *Coll. Czch. Chem. Comm.* 32, 56(1967).

26. E. Eisenberg, C.W. Tobias, C.R. Wilke, *J. Electrochem. Soc.* 101, 306(1954).

27. D.J. Robinson, D.R. Gabe, *Trans. Inst. Met. Fin.* 48, 35(1970).

28. D.R. Gabe, D.J. Robinson, *Trans. Inst. Met. Fin.* 49, 17(1971).

29. J. Newman, *Electrochemical Systems* (New York, NY: Prentice Hall, 1973).

30. J. Newman, *Ind. Eng. Chem.* 60, 12(1968).

31. A. Kappesser, I. Cornet, R. Grief, *J. Electrochem. Soc.* 118, 1957(1971).

32. F. Coeuret, J. Legrand, *Electrochimica Acta* 26, 865(1981).

33. E. Brunner, *Z. Phys. Chem.* 47, 56(1904); 51, 95(1905); 58, 1(1907).

34. C.W. Bennett, *Trans. Electrochem. Soc.* 21, 245(1912).

35. D.D. Macdonald, B.C. Syrett, S.S. Wing, *Corrosion* 34, 289(1978).

36. M. Stern, A.L. Geary, *J. Electrochem. Soc.* 104, 56(1957).

37. B.C. Syrett, D.D. Macdonald, *Corrosion* 35, 505(1979).

38. L. Giuliani, M.G. Bombara, *Brit. Corrosion J.* 5, 179(1970).

39. L. Giuliani, A. Tamba, C. Modena, *Corrosion Sci.* 11, 485(1971).

40. R.R. North, M.J. Pryor, *Corrosion Sci.* 10, 297(1970).

41. R.G. Blundy, M.J. Pryor, *Corrosion Sci.* 12, 65(1972).

42. J.M. Popplewell, R.J. Hart, J.A. Ford, *Corrosion Sci.* 13, 295(1973).

43. J.G. Knudsen, *Amer. Inst. Chem. Engrs. J.* 8, 565(1962).

44. D.D. Macdonald, *J. Electrochem. Soc.* 125, 1443(1978).

APPENDIX

A

CONVERSION OF CORROSION RATE FROM $1/R_p$(OHMS^{-1}) TO MPY

- According to the Stern-Geary treatment of electrochemical corrosion, the corrosion current (I_{corr}) is proportional to the reciprocal of the polarization resistance as shown in the following equation:

$$I_{corr} = \frac{B_a B_c}{2.303 R_p (B_a + B_c)} \tag{A.1}$$

Equation (A.1) may be integrated over time (t) to yield the charge (Q_{corr}) associated with metal loss,

$$Q_{corr} = \left[\frac{B_a B_c}{2.303 (B_a + B_c)}\right] \int_o^t \left(\frac{1}{R_p}\right) dt = \frac{BA_t}{2.303} \tag{A.2}$$

in which $\quad B = \dfrac{B_a B_c}{(B_a + B_c) A_t} = \displaystyle\int_o^t \left(\frac{1}{R_p}\right) dt$

Application of Faraday's law then yields the weight loss (ΔW) in units of grams as:

$$\Delta W = \frac{MW_{corr}}{nF} = \frac{MBA_t}{2.303\,nF} = \left(\frac{MB}{2.303\,nF}\right)\int_o^t \left(\frac{1}{R_p}\right) dt \tag{A.3}$$

where M and n are the composition-averaged atomic weight and change in oxidation state, respectively. M and n are given by Equations (A.4) and (A.5) in terms of the mole fractions of copper (X_{Cu}) and nickel (X_{Ni}), and the molecular weight of copper (M_{Cu}) and nickel (M_{Ni}):

$$M = X_{Cu}M_{Cu} + X_{Ni}M_{Ni} \qquad (A.4)$$

$$n = X_{Cu} + 2X_{Ni} \qquad (A.5)$$

It is assumed that the copper and nickel components are oxidized to the +1 (cuprous) and +2 (nickelous) states, respectively. For 90:10 Cu-Ni alloy, M is 63.04 g/mole and n is 1.107 g/mole. Substituting these values into Equation (A.3) gives:

$$\Delta W = 2.563 \cdot 10^{-4} BA_t$$

for 90:10 Cu-Ni alloy. The above equation was used to calculate the weight loss in this work.

● Differentiation of Equation (A.3) with respect to t gives the weight loss in units of grams/s:

$$\frac{d\Delta W}{dt} = \left(\frac{MB}{2.303nF}\right)\frac{1}{R_p} \qquad (A.6)$$

This can be converted to corrosion rate in units of mpy by:

$$mpy = \frac{d\Delta W}{dt} \cdot \frac{86,400 \cdot 365}{1}\left(\frac{sec}{year}\right) \cdot \frac{1}{\rho}\left(\frac{cm^3}{g}\right) \cdot \frac{1}{A}\left(\frac{1}{cm^2}\right) \qquad (A.7)$$

where ρ is the density of the 90:10 Cu-Ni alloy, and A is the exposed area.

Substituting appropriate values (assuming $B = 0.07$ volt) into Equations (A.6) and (A.7) yields the conversion equations for corrosion rate from $1/R_p$ (ohms^{-1}) to mpy:

For the flow-through pipe channel—

$$mpy = 3{,}958.4\,(mpy \cdot ohms^{1}) \cdot \frac{1}{R_p}\,(ohms^{-1})$$

For the annular flow channel—

$$mpy = 3{,}272.3\,(mpy \cdot ohms^{1}) \cdot \frac{1}{R_p}\,(ohms^{-1})$$

For the rotating cylinder system—

$$mpy = 3{,}272.3\,(mpy \cdot ohms^{1}) \cdot \frac{1}{R_p}\,(ohms^{-1})$$

NOMENCLATURE

(The number in parenthesis is the corresponding value for the systems studied.)

A = Electrode surface area, cm^2
    A = 6.283 cm^2 for the flow-through pipe channel
    A = 7.601 cm^2 for the annular flow channel and the rotating cylinder system

C_b = Bulk concentration of transferred species, $mole/cm^3$
    C_b = 8 ppm $[O_2]$ = $2.925 \cdot 10^{-7}$ mole O_2/cm^3 under air-saturated conditions

D = Diffusivity, cm^2/s
Diffusivity of oxygen in 1 M NaCl solution = $2.3 \cdot 10^{-5}$ cm^2/s

$E_{corrosion}$ = Corrosion potential, mV

F = Faraday's constant

I_ℓ = Diffusion-limited current, mA
    I_{ℓ_1} = first plateau in the cathodic polarization curve
    I_{ℓ_2} = second plateau in the cathodic polarization curve

i_ℓ = Averaged steady state diffusion-limited current, mA
    i_{ℓ_1} = value of I_{ℓ_1} at the first plateau in the cathodic polarization curve
    i_{ℓ_2} = value of I_{ℓ_2} at the second plateau in the cathodic polarization curve
    K_1 = Overall mass transfer coefficient calculated from i_{ℓ_1}, cm/s
    K_2 = Mass transfer coefficient calculated from i_{ℓ_2}, cm/s
    K_f = Mass transfer coefficient in the surface film, cm/s
    K_h = Mass transfer coefficient in the diffusion boundary layer calculated from previously published mass transfer rate equation, cm/s

Electrode length, cm (L = 1.27 cm for all hydrodynamic systems)

M = Composition averaged atomic weight (M = 63.04 g/mole)

n   =   Composition averaged change in oxidation state (n = 1.107 g/mole)

n   =   Number of electrons involved in reaction (n = 4 for oxygen reduction)

U   =   Peripheral velocity of the rotating electrode, m/s

V   =   Average linear flow velocity, m/s

ΔW =   Weight loss, g

X   =   Inside diameter of the specimen for the flow-through pipe channel (= 1.575 cm), or effective diameter for the annular flow channel (= 0.635 cm), or outside diameter of the specimen for the rotating cylinder system (= 1.905 cm)

α =   Annulus radius ratio, r_1/r_2 = 0.9525 cm/1.27cm = 0.75

R_1 =   Radius of the inner rotating electrode for the rotating cylinder system, cm

R_2 =   Inside radius of the stationary counter electrode for the rotating cylinder system, cm

B_a =   Anodic Tafel slope, volts

B_c =   Cathodic Tafel slope, volts

B' =   Velocity gradient at conduit walls, S^{-1}

λ =   Ratio of radial distance of point of maximum velocity to outer radius (λ^2 = 0.7604)

μ =   Viscosity, g/s · cm (μ = 1.1894 g/s · cm for 1 M NaCl solution)

ρ =   Density of solution, g/cm^3 (ρ = 1.17g/cm^3 for 1 M NaCl solution)

ν =   Kinematic viscosity, cm^2/s (ν = 1.01656 · $10^{-2}cm^2/s$ for 1 M NaCl solution)

Re   =   Reynolds number = XV_t/μ or $\rho XU/\mu$

Sc   =   Schmidt number = $\mu/\rho D$ (Sc = 442 for 1 M NaCl solution at 25°C [77°F])

Sh   =   Sherwood number = Kx/D ($Sh_1 = K_1X/D$, $Sh_2 = K_2X/D$, $Sh_h = K_hX/D$)